41

Springer Series in Chemical Physics

Edited by Koichi Shimoda

Springer Series in Chemical Physics
Editors: V. I. Goldanskii R. Gomer F. P. Schäfer J. P. Toennies

Volume 40 **High Resolution Spectroscopy in Transient Molecules**
Editor: E. Hirota

Volume 41 **High Resolution Spectral Atlas of Nitrogen Dioxide 559–597 nm**
By K. Uehara and H. Sasada

Volumes 1–39 are listed on the back inside cover

K. Uehara H. Sasada

High Resolution Spectral Atlas of Nitrogen Dioxide 559–597 nm

With 7 Figures and 179 Charts

Springer-Verlag
Berlin Heidelberg New York Tokyo

Kiyoji Uehara, Ph. D.
Hiroyuki Sasada, Ph. D.

Faculty of Science and Technology, Keio University,
3-14-1 Hiyoshi, Kohoku-ku, Yokohama 223, Japan

Guest Editor:

Professor Koichi Shimoda

Faculty of Science and Technology, Keio University,
3-14-1 Hiyoshi, Kohoku-ku, Yokohama 223, Japan

Series Editors

ISBN-13: 978-3-642-82411-1 e-ISBN-13: 978-3-642-82409-8
DOI: 10.1007/978-3-642-82409-8

Softcover reprint of the hardcover 1st edition 1985

Bookbinding: J. Schäffer OHG, 6718 Grünstadt
2153/3130-543210

Preface

Nitrogen dioxide (NO_2) is notorious for its complex visible spectrum and has frustrated many chemists and physicists. Despite their intense investigations the molecule still resists complete analysis of its spectrum. Because of the high density of lines, it is apparent that high-resolution measurements are essential to a thorough understanding of the spectrum.

The major part of this book consists of an atlas, described in Chap. 2, of the absorption spectrum and the Stark modulation spectrum of NO_2 measured using a cw dye laser. The narrow spectral width as well as the wide scan range of the dye laser made it possible to record the spectra continuously over a wide range with Doppler-limited resolution. In the spectral range 16751–17885 cm^{-1}, about 13300 peaks have been observed in the absorption spectrum and their wave numbers, calibrated against Ar lines, are listed in the tables. The Stark modulation spectrum also shown provides useful information about the energy level structures complementary to that obtained from the absorption spectrum.

Although understanding of the whole visible spectrum of NO_2 is far from complete, some selected parts of it have been successfully analyzed by various experimental methods, especially those using lasers. Radiative and coherence properties, concerning the excited-state dynamics of NO_2, have also been unraveled using lasers. Recent progress in laser spectroscopic studies of this molecule is reviewed briefly in Chap. 3. In Chap. 4 a selected bibliography of recent (1978 onwards) studies, including non-laser experiments and theoretical works, is given.

Nitrogen dioxide is one of the key molecules in the research of the atmosphere and combustion products. It is our hope that this volume will contribute to the progress of studies both in spectroscopy and in related fields of application.

It is our pleasure to thank several people who helped in the experiment and the preparation of the manuscript. Mr. Takashi Iwamoto assisted effectively in the early stage of the data analysis. Dr. Yoshifumi Ueda's advice about the experimental techniques is appreciated. Dr. Nobuo Takeuchi kindly provided some gas samples and also supplied us with information about literature on NO_2, including their unpublished work. Discussions with Dr. Hans G. Weber concerning the present publication plan were valuable. Tsukiko Uehara helped her father voluntarily in the reproduction of the charts. Mrs. Chitose Hikami typed the drafts beautifully and speedily. We wish to express our special appreciation to Professor Koichi Shimoda for his persistent encouragement throughout the work and careful reading of the manuscript. Finally we would like to thank Dr. H. Lotsch and coworkers of Springer-Verlag for help and advice throughout the preparation and production period of this book.

Yokohama, Japan
December 1984

Kiyoji Uehara
Hiroyuki Sasada

Contents

1. Introduction

Nitrogen dioxide (NO_2) shows a much more complicated visible absorption spectrum than expected for a triatomic molecule. Fine structure due to the spin $S = 1/2$ of the unpaired electron, hyperfine structure due to the nuclear spin $I = 1$ of the ^{14}N atom, and overlapping of the excited electronic states are obviously among the causes of its complexity. In addition, it has been found that the coupling between the electronic states through various perturbation mechanisms plays an important role in its irregular spectrum. Quite a few researchers have challenged the complicated spectrum hoping for better understanding of the physics of this frustrating molecule. Despite their efforts, most portions of the spectrum remain to be analyzed.

The visible absorption lines of NO_2 are so dense that a very high resolution is necessary to separate single vibration-rotation lines. Although partial analysis of the NO_2 spectrum has been successful in some selected bands, wide-range, high resolution measurements are still essential for a thorough understanding of the spectrum. The spectral atlas of NO_2 by *Hsu* et al. [1.1] published in 1978 has been up to now the most extensive data source to meet this need. Their atlas consists of the densitometer tracings of the absorption spectrum photographed with a grating spectrograph. Wavelengths and wave numbers are given for prominent lines. The book also contains a good review and a summary of relevant studies as well as a list of publications up to 1977. Some other data sources for the wide-range spectrum of NO_2 are described there. It is apparent, however, that still higher resolution data would be very useful.

In recent years various new techniques using lasers have been introduced to spectroscopy in addition to, or in conjunction with, conventional techniques. These new techniques have indeed made a number of precise measurements possible for some particular transitions in narrow spectral ranges. However, no extensive application of lasers to high resolution measurements of atomic or molecular spectra in a very wide range ($\gtrsim$1000 cm^{-1}) has been reported, probably with the exception of a work by *Smalley* et al. [1.2], see also Sect. 3.2. It seems quite natural to make use of tunable lasers for collecting spectroscopic data of NO_2 over a wide range.

This book presents, following Chap. 4, a high resolution spectral atlas of NO_2 measured with a wide-scan single-mode dye laser. Both the absorption spectrum and the Stark modulation spectrum are given with a resolution limited by the Doppler width. The spectral coverage in this measurement is 559–597 nm (16 751–17 885 cm^{-1}), which is shown in Fig. 1.1, along with the region covered by the atlas of *Hsu* et al. [1.1], the region investigated

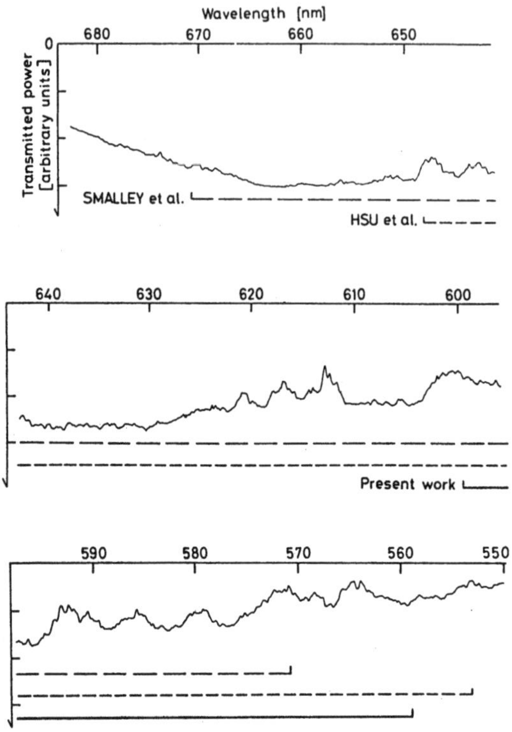

Fig. 1.1. Spectral range covered by the present atlas. The ranges studied by *Hsu* et al. [1.1] and by *Smalley* et al. [1.2] are also shown. The curve represents the low resolution absorption spectrum of NO_2 measured using a monochromator. No correction was made for wavelength dependences of the radiation power of the source (tungsten filament lamp), the efficiency of the grating, the sensitivity of the photomultiplier, etc. Absorption path length 10 cm; pressure 20 Torr; resolution 0.08 nm

by *Smalley* et al. [1.2], and a low resolution absorption spectrum of NO_2. About 13300 lines are resolved and their wave numbers listed in the present atlas.

In addition to high resolution and high signal-to-noise ratio, this atlas features the Stark modulation spectrum, recorded simultaneously with the absorption spectrum. Useful information for line assignment can be extracted from the Stark modulation spectrum, and interpretations of the observed spectrum are discussed in Sect. 2.4.

The use of lasers in spectroscopy has become more and more widespread and important. While analysis of the whole spectrum of NO_2 is far from complete, some selected parts have been analyzed by means of laser-induced fluorescence, laser excitation spectroscopy, double resonance laser spectroscopy, and so forth. Also, the radiative and coherence properties, concerning the excited-state dynamics of this molecule, have been investigated by lasers. In Chap. 3, a brief review of recent spectroscopic studies on NO_2 using lasers is given. In Chap. 4, selected publications from 1978 onwards, including those describing non-laser experiments and theoretical work, are listed.

Nitrogen dioxide is one of the key molecules in atmospheric research. *Wilkerson* et al. [1.3] measured the absorption spectrum of NO_2 under ambient conditions with a 0.003-nm resolution. The densitometer tracings of the spectrum in the 405- to 515-nm region are given in their report. On the other hand, data on about 10000 lines of NO_2 in the infrared to radiowave regions is included in the AFGL (Air Force Geophysics Laboratory) atmospheric absorption line parameters compilation [1.4,5]. Line positions, intensities, half-widths, and assignments are compiled there.

2. High Resolution Spectral Atlas of NO_2 Between 559 and 597 nm

The spectral atlas following Chap. 4 consists of the absorption spectrum and the Stark modulation spectrum of NO_2, divided into 179 charts, along with the wave-number tables of the observed absorption peaks. The cw dye laser used as a light source had a very narrow spectral width, so that the observed spectra are essentially Doppler-limited. Argon absorption lines were used for wave-number calibration. The spectral interval covered by the measurements is 559–597 nm (16 751–17 885 cm^{-1}), which is the major part of the tuning range of a Rhodamine 6G dye laser. Each of the 179 sets of traces was recorded in a single scan over a frequency range of $\cong$7 cm^{-1}. The total number of peaks listed in the tables is about 13 300.

The absorption spectrum was obtained by monitoring the intensity of the dye-laser beam transmitted through an absorption cell, while the Stark signal was provided by lock-in detection of the effect of the Stark modulation field on the absorption. These spectra were recorded simultaneously on a chart recorder. In spite of the simplicity of the experimental arrangement, a very high signal-to-noise ratio was achieved both in the absorption spectrum and in the Stark modulation spectrum. This was due to the frequency and power stabilization of the dye laser and to the differential detection technique described below.

The frequency coverage presented is about 40% of that in the spectral atlas of *Hsu* et al. [2.1]. However, the average line density of 12/cm^{-1} in our tables is more than twice their value. This arises from the differences in resolution and signal-to-noise ratio between the two measurements. In our atlas, furthermore, all absorption lines whose wave numbers were measured are marked and numbered on each trace so that the reader can find their wave numbers in the table immediately.

All experiments needed to make the atlas were carried out in 1977 and 1978 by Uehara (then at the University of Tokyo). Computer-assisted position reading of the absorption lines and tabulation of the wave numbers were done by Sasada.

Experimental details are given in Sect. 2.1, and in Sect. 2.2 the method of frequency calibration is described. Those readers who are interested only in the charts themselves can skip to Sect. 2.3 where they are described. In Sect. 2.4, interpretations and a partial analysis of the Stark modulation spectrum are given.

2.1 Experimental Details

A schematic of the experimental setup is shown in Fig. 2.1.

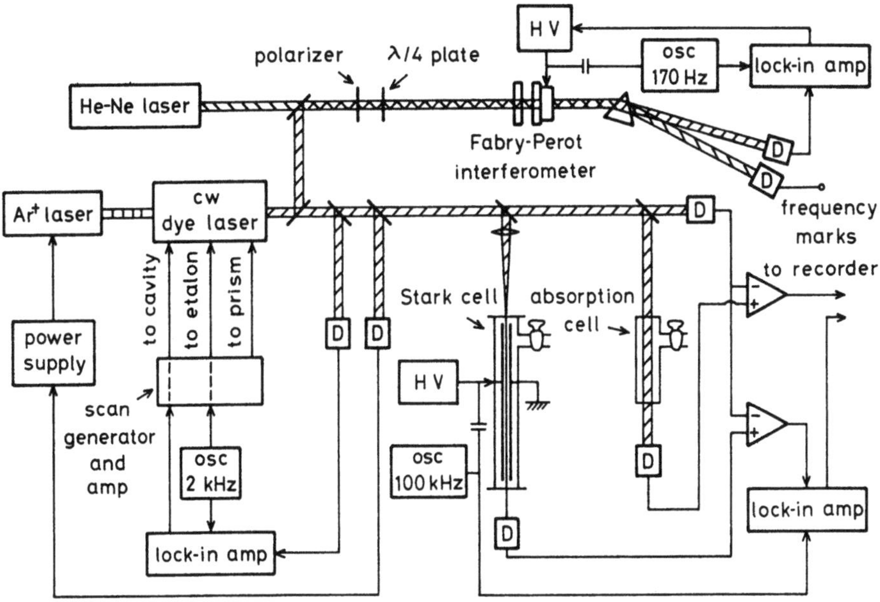

Fig. 2.1. Schematic drawing of the experimental arrangement. (*D*) detector, (*HV*) high voltage power supply

2.1.1 Dye Laser

For measurements over such a wide spectral range as the one we present, it is desirable that the laser be widely tunable in a single scan. In the measurements a commercial dye laser (Spectra Physics, Model 580, flowing dye cell type) was modified by adding two homemade servo loops so as to be tunable over as wide a range as 14 cm^{-1}, while retaining a narrow spectral width of 20 MHz [2.2]. The first servo loop prevents mode jumps by locking the cavity to an internal etalon which is modulated at 2 kHz; the error signal is provided by lock-in detection of the resulting amplitude modulation of the output power. The second servo loop stabilizes the dye-laser power by controlling the pumping power delivered by an argon-ion laser (Spectra Physics, Model 165). The output current of the photodiode detector monitoring the dye-laser power is fed into the "light control" circuit in the commercial power supply of the argon-ion laser. Power fluctuations of the dye laser are thus stabilized to be less than 1% when measured with a time constant of 10 ms. A change in absorption as small as 0.3% can be detected by the differential detection technique, as mentioned below. Recently, we succeeded in wide scanning of a dye laser over 100 cm^{-1} by a similar technique [2.3]. A computer-controlled wide-scan dye-laser system is now also available on the market [2.4].

In our system, stable output powers up to 30 mW are obtained at any wavelength between 555 and 600 nm from a Rhodamine 6G-methanol solution excited by no more than 1 W of the 514.5-nm line. Because of the nonlinear response of the piezoelectric transducer used in the internal etalon, the frequency scan rate is not linear. Therefore, the frequency marks of a high finesse interferometer are used for frequency interpolation as described in Sect. 2.2.

2.1.2 Absorption Spectrum

The absorption cell is a Pyrex tube 10 cm long, filled with the NO_2 sample at a pressure of 9.5 Torr. The dye-laser beam is divided into two parts by a beam splitter: one is focused on a photodiode detector after passing through the absorption cell and the other is detected directly by another photodiode. The beam incident on the absorption cell has a spot of $\cong$1 mm and its power is typically $\cong 100\,\mu$W. The output current of the two detectors are fed into a differential amplifier which is adjusted so as to give null output when the absorption cell is replaced by an empty cell. The output of the amplifier is recorded on a chart recorder without the use of a filtering circuit. The typical noise level in this differential detection system is shown in Fig. 2.2a together with some absorption signals. The frequency sweep rate is $\cong 1\ \mathrm{cm}^{-1}\ \mathrm{min}^{-1}$. The vertical scale represents a quantity $(I_0 - I)/I_0$, where I_0 and I are the intensities of the laser beam before and after passing through the absorption cell, respectively. The minimum detectable absorption change as estimated

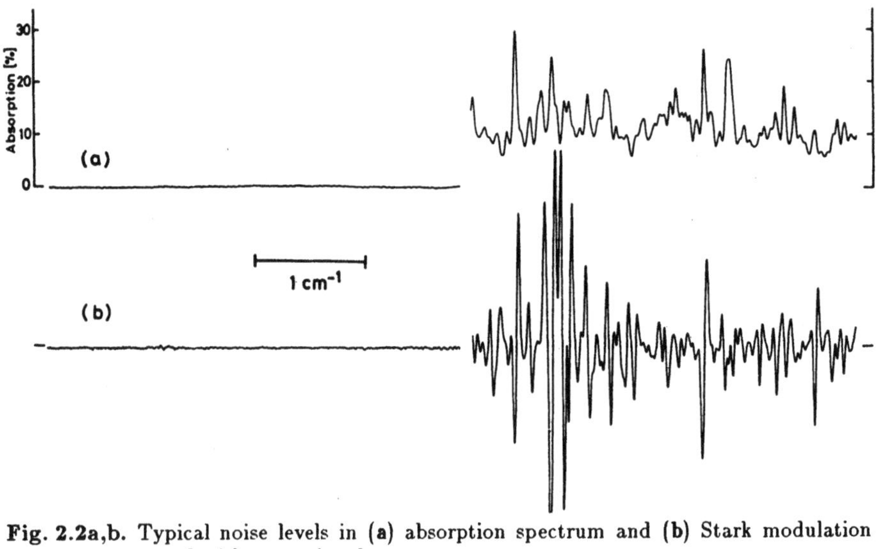

Fig. 2.2a,b. Typical noise levels in **(a)** absorption spectrum and **(b)** Stark modulation spectrum compared with some signals

from the noise level is 0.3%. Sometimes, a drift of the zero absorption level occurs owing to the deflection of the laser beam during a scan. This drift results in an ambiguity of up to ±2% in the vertical scale. However, the drift is slowly varying and therefore does not affect the minimum detectable absorption change.

2.1.3 Stark Modulation Spectrum

The Stark absorption cell is composed of a pair of Stark electrodes supported in a Pyrex tube. The electrodes of 0.5-mm spacing are made of silvered glass plates 90 cm in length and 2 cm in width. A focussing lens of $f = 80$ cm is used to couple the laser beam to a narrow gap between the electrodes. A sinusoidal Stark voltage at 100 kHz is superposed on a dc voltage. The resulting change at 100 kHz in the power of the transmitted beam is detected by a lock-in amplifier and recorded on the chart recorder simultaneously with the absorption spectrum. Again, a differential amplifier like that mentioned above is used before the lock-in amplifier. This greatly reduces the 100-kHz noise component originating from the dye laser. Figure 2.2b represents the noise level together with some signals in the Stark modulation spectrum. The dc field is 12 kV/cm and the modulation field is 7 kV/cm peak-to-peak. The Stark field is perpendicular to the optical electric field of the incident laser beam throughout the measurement.

The Stark absorption cell is filled with the sample gas to a pressure of 1.0 Torr. However, the pressure decreases with time because NO_2 is chemically very active with the metallic surface of the electrodes. Therefore, the gas is added or refilled after each ten scans or so, to keep the pressure between 1.0 and 0.5 Torr.

2.1.4 NO_2 Samples

The NO_2 gas used in the measurement was prepared by mixing NO and O_2, both 99.9% pure and of natural isotopic abundance. They were introduced into a Pyrex vessel with an excess of O_2. The mixture at once turned brown, the color of NO_2. Then, the gas was frozen in a cold trap into white condensate with bluish impurities. After the remaining gas was pumped out, the condensate was vaporized by raising the temperature. This condensation-vaporization cycle was repeated several times until pure colorless condensate was obtained.

2.2 Wave-Number Calibration

The tables in the atlas give the vacuum wave numbers [cm^{-1}] of the peaks in the absorption spectrum. Wave-number calibration by means of Ar I lines is accomplished in the following three steps.

In the first step, simultaneous recording of the NO_2 absorption spectrum and the frequency marks of a Fabry-Perot interferometer is made over the whole spectral range covered.

As shown in Fig. 2.1, a Fabry-Perot interferometer with a 1.6-mm air space is servo-controlled so that the 632.8-nm He-Ne laser line coincides with one of the transmission maxima of the interferometer. Although the He-Ne laser oscillates on a few longitudinal modes with 500-MHz spacings and is not frequency-stabilized, the power spectrum of the laser viewed through the low resolution interferometer has a single peak which is stable within ±90 MHz (±0.003 cm^{-1}). The frequency marks are then provided by the transmission of a dye-laser beam, which is incident on the interferometer collinearly with the He-Ne-laser beam. A prism is used to separate the two beams behind the interferometer. The dye laser is swept over the full 14-cm^{-1} scan range so as to give four or five frequency marks, which make accurate interpolation possible. Figure 2.3 represents an example of the simultaneous recording of frequency marks and the NO_2 absorption lines. The reproducibility of the frequency marks was found to be typically within ±0.006 cm^{-1}; this number is limited by the deflection of the dye-laser beam incident on the interferometer and the instability of the reference He-Ne laser.

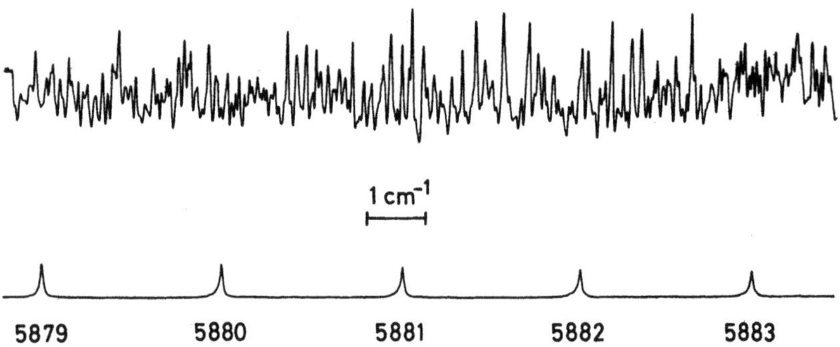

Fig. 2.3. An example of simultaneous recording of the NO_2 absorption spectrum and the frequency marks around 565 nm. N is the absolute order number of interference

Order numbers N^* are then given to the frequency marks starting at the long-wavelength side. To conserve a definite relation between the order numbers and corresponding frequencies, it is essential to keep the interferometer on the same transmission peak of the He-Ne-laser line throughout all such measurements. The same transmission peak is easily found by observing two transmission curves for the 632.8-nm He-Ne-laser line and the 514.5-nm argon-ion-laser line simultaneously. The absolute order number of interference N is given by $N = N^* + m$, where m is a fixed integer.

The second step is to determine an explicit relationship between the order numbers and wave numbers.

The resonance condition of an interferometer is given by

$$\frac{2L}{\lambda_{\text{air}}} + \frac{\phi_\lambda}{\pi} = N, \tag{2.1}$$

where L is the spatial length of the interferometer, λ_{air} the air wavelength of the incident beam, and ϕ_λ the phase shift at the mirror surfaces. Since the wavelength changes significantly, the wavelength dependence of ϕ_λ must be taken into account. The mirrors used here have thirteen dielectric layers of SiO_2 and TiO_2 alternately deposited on a quartz substrate and have a high reflectance (over 99.5%) throughout the 555- to 600-nm range. A theoretical calculation shows that ϕ_λ is approximated well in this spectral range by a linear function of $1/\lambda_{\text{air}}$, i.e.,

$$\phi_\lambda = \pi(a/\lambda_{\text{air}} + b), \tag{2.2}$$

where a and b are constants, and $|b| \ll 1$.

From (2.1,2) we obtain a linear relation between $1/\lambda_{\text{air}}$ and N:

$$\frac{1}{\lambda_{\text{air}}} = \frac{N - b}{2L + a}. \tag{2.3}$$

This relation can be extended to any wavelength if N is allowed to take non-integral values. Thanks to the condition $|b| \ll 1$, we can determine the two independent unknown parameters b and $2L + a$ in this equation if we know the order numbers N^* (not necessarily integers) for at least two wavelengths. Note also that their absolute order numbers N can be determined.

Four argon absorption lines are used as wavelength references. A discharge is produced in argon gas at 0.5 Torr in a Pyrex tube of length 30 cm and diameter 5 mm by a dc current of 25 mA. Figure 2.4 shows the argon absorption line at 565 nm, recorded simultaneously with the NO_2 absorption lines which also appeared in Fig. 2.3. The order number for this argon line is then obtained from these figures by interpolation. Four sets of values of λ_{air} and N for four argon lines are used to determine the unknown

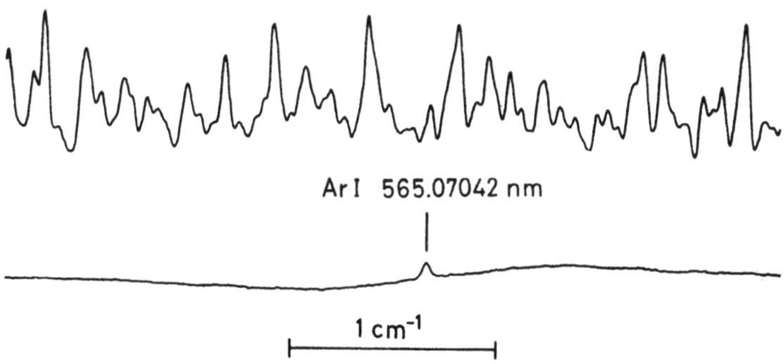

Fig. 2.4. Simultaneous recording of the 565-nm Ar I reference line and the NO_2 absorption lines

parameters in (2.3) with a least-squares fit. The results are $b = 0.2295$ and $2L + b = 0.33234172$ cm. The wavelengths of the four Ar I lines [2.5] and residual errors in the fit are as follows:

λ_{air} [nm]	$\Delta(1/\lambda_{air})$ [cm^{-1}]
560.67328	−0.004
565.07042	+0.004
591.20848	+0.004
592.88119	−0.005

standard deviation = 0.005 cm^{-1}

Another equation we need in order to calculate the vacuum wave number ν is

$$\nu = \frac{1}{n'_\lambda \lambda_{air}}, \tag{2.4}$$

where n'_λ is the refractive index of air at wavelength λ in air. n'_λ can be replaced, to a good approximation, by n_λ, the refractive index of air at the vacuum wavelength λ. We use here an Edlén expression for n_λ of standard air:

$$(n_\lambda - 1) \times 10^8 = 6432.8 + 2949810/(146 - \lambda^{-2}) + 25540/(41 - \lambda^{-2}), \tag{2.5}$$

where λ is the vacuum wavelength in μm. A correction due to the difference in n_λ between standard air and the real atmosphere in the laboratory is negligible because most of this difference is compensated for by the correction for the the He-Ne-laser line used to stabilize the interferometer.

In the final step, the wave number for each peak in the NO_2 absorption spectrum is determined by interpolation between the calibrated frequency marks.

The position x of the absorption peaks and frequency marks on a chart is read with a resolution of 5 μm (3×10^{-5}-cm^{-1} wave-number resolution) by a position reader (Sony Magnescale, Model S-1), which provides a digital output to be sent to a computer. A polynomial relationship between ν and x is assumed for interpolation:

$$\nu = a_0 + a_1 x + a_2 x^2 + a_3 x^3 + a_4 x^4. \tag{2.6}$$

Firstly, positions of five frequency marks are read in order to determine the unknown parameters $a_0 - a_4$. When only four frequency marks are present, a_4 is set equal to zero. Then, the position of each absorption peak is read and the wave number is calculated from (2.6). The cursor of the Magnescale is operated manually, but all computations are performed by a microcomputer (DEC, Model LSI-11-16J).

Because the NO_2 spectrum is very dense, it often happens that an absorption line is appreciably deformed by a neighboring line or lines. Therefore, we read in all cases the positions of peaks in the absorption spectrum instead of reading line centers. In the spectral atlas, all peaks measured are marked by dots and their serial numbers are given every ten dots. The wave-number measurement is repeated for all peaks and the average of two measurements is given in the tables. Normally, the two measurements coincide within $\pm 0.002\,cm^{-1}$.

The accuracy of interpolation depends on how well the assumed formula (2.6) fits the real $\nu - x$ curve. Wave-number measurements of the same absorption lines using different sets of frequency marks showed that the deviation of the polynomial from the real curve is within $\pm 0.004\ cm^{-1}$. The absolute accuracy of calibration is then estimated to be $\pm 0.015\,cm^{-1}$, the sum of three errors: the reproducibility of the frequency marks, the standard deviation in the fit of the reference lines, and the accuracy of interpolation.

2.3 Using the Atlas

In each of the following charts, the upper tracer is the absorption spectrum while the lower is the Stark modulation spectrum. The curves were reproduced photographically from the original chart recorder tracings. Vertical lines between the curves represent wave-number scales at 1-cm^{-1} intervals. Vacuum wave numbers [cm^{-1}] for dotted peaks (not line centers) in the absorption spectrum are given with their serial numbers.

Various parameters needed for reading the atlas are summarized here:

Dye laser
Spectral width: 20 MHz
Maximum scan range: 14 cm^{-1}
Scan rate: $\cong 1\ cm^{-1}min^{-1}$

Absorption spectrum
Absorption path length: 10 cm
Sample pressure: 9.5 Torr
Incident laser power: $\cong 100\ \mu W$
Laser spot size: $\cong 1$ mm
Accuracy of vertical scale: $\pm 2\%$
Noise level: see Fig. 2.2a

Stark modulation spectrum
Absorption path length: 90 cm
Sample pressure: 0.5–1.0 Torr
Electrode spacing: 0.5 mm
dc Stark field: 12 kV/cm
ac Stark field: 7 kV/cm peak-to-peak, sinusoidal
Stark field direction: perpendicular to optical electric field
Noise level: see Fig. 2.2b

Wave-number calibration
Wavelength references: Ar I absorption lines
Absolute accuracy: $\pm 0.015\ cm^{-1}$

2.4 Remarks on the Stark Modulation Spectrum

The Stark modulation technique is a sensitive method of detecting absorption lines of polar molecules and has been very widely used in microwave spectroscopy and recently also in infrared laser spectroscopy. At the same time, the Stark effect of a molecule provides useful information in the identification of its absorption lines.

Until ten years ago only a few researchers had applied the Stark modulation technique to visible and uv spectroscopy [2.6], probably for lack of good radiation sources with a narrow spectral width. Recently, however, its usefulness in dye-laser spectroscopy has been demonstrated [2.2,7,8]. Usually, lines with low rotational quantum numbers are enhanced in the Stark modulation spectrum, as exemplified by those of ICl [2.8]. In spectra thus simplified one band can be separated from other overlapping bands and the rotational structure analyzed very easily.

At first, a similar simplification of the spectrum was expected for NO_2. Contrary to expectations, however, the observed Stark modulation spectrum showed a very dense and complicated structure. As can be seen in the atlas, almost all absorption lines suffer a sizable Stark effect. Even line shapes showing a first-order Stark shift appear rather frequently.

Generally speaking, an appreciable amount of Stark shift can be expected for a level only when there is a nearby level which is connected to it by an electric-dipole matrix element. The first-order Stark shift requires a coincidence of such levels. The observed high density of lines in the Stark modulation spectrum means that such coincidences or near-coincidences are quite common. This is unexpected, because NO_2 is a near-prolate asymmetric-top molecule with a permanent dipole moment directed along the *b*-axis and therefore the dipole-connected levels in the same vibronic state come close to each other only accidentally. Furthermore, one component of the K-type doublet levels of $N^{16}O_2$ is missing due to the spin statistics for the O nuclei.

A rough estimate, assuming harmonic vibrational potentials in the ground and excited electronic states, gives a vibronic state density of $0.15/cm^{-1}$ at $17\,500\ cm^{-1}$ above the ground state. This number seems too small to explain the frequent accidental coincidence of rovibronic levels.

One possible interpretation of the richness of the Stark signals in NO_2 is suggested by the unequal N-O bond lengths in the exicted state pointed out by several authors [2.9–14]. This asymmetric geometry leads to doublet structure in each rotational level. The two sublevels are dipole-connected to each other through tunneling between the two equilibrium asymmetric geometries. If the doubling is as small as a few cm^{-1} almost all levels show Stark shifts of a detectable amount and the first-order Stark shifts are more likely to occur.

Another interpretation is that most of the Stark signals originate from asymmetric isotopic species such as $^{16}ON^{18}O$. The inequality of the two O nuclei makes K-type doubling allowed. However, this explanation seems less reasonable, because the natural abundance of ^{18}O is only 0.2% and because the a-component of the permanent dipole moment is very small.

Despite the complexity of the Stark modulation spectrum, useful information for line assignment can be derived from it. Some of the Stark patterns have been clearly explained. A calculation of the energies of the low-lying rotational levels in the ground vibronic state shows that the 7_{17} level is located above the 8_{08} level by only $0.5\,cm^{-1}$. The two levels are dipole-connected. Other pairs of levels have a much greater separation or are dipole-unconnected. It is expected therefore that transitions starting from the 7_{17} or 8_{08} levels give rise to prominent Stark signals, and actually, many lines in the Stark modulation spectrum have been assigned to such transitions. In Figs.2.5,6 some of these transitions in the 593.3- and 585.1-nm bands are shown. The 593.3-nm band has been extensively analyzed by laser-induced fluorescence [2.15], by microwave-optical double resonance [2.16], and by the Zeeman effect [2.17]. Some of the low-N transitions in the 585.1-nm band have been identified in the excitation spectrum of cooled molecules [2.18]. The upper electronic state of both band is the $^2B_2$ state. The doublet structure of each line is primarily due to the large spin splitting in the upper state.

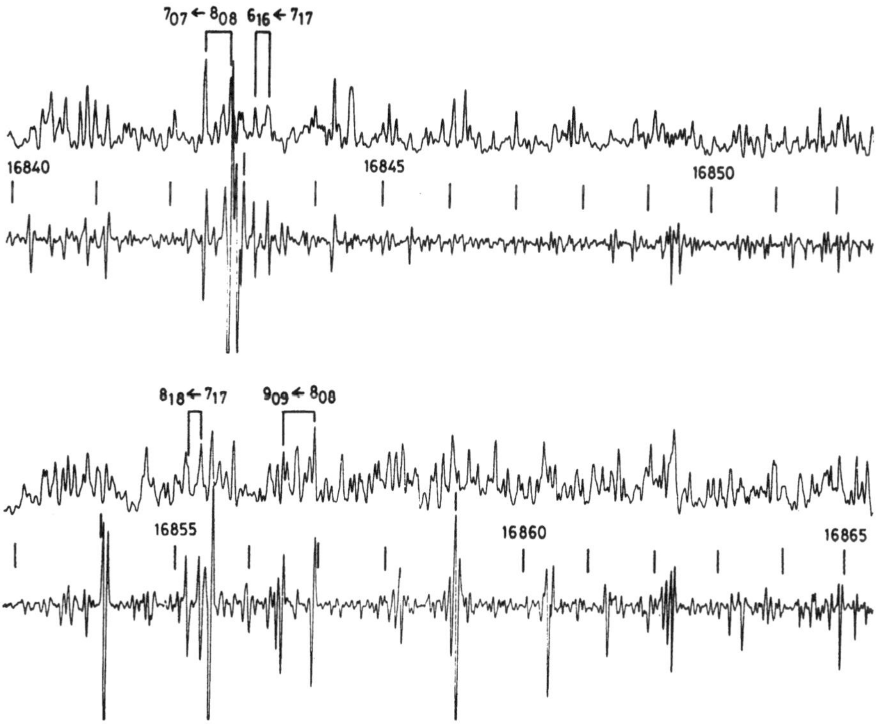

Fig. 2.5. The absorption spectrum (*upper trace*) and the Stark modulation spectrum (*lower trace*) of NO_2 around 593.3 nm. The frequency marks are given at $1\,cm^{-1}$ intervals

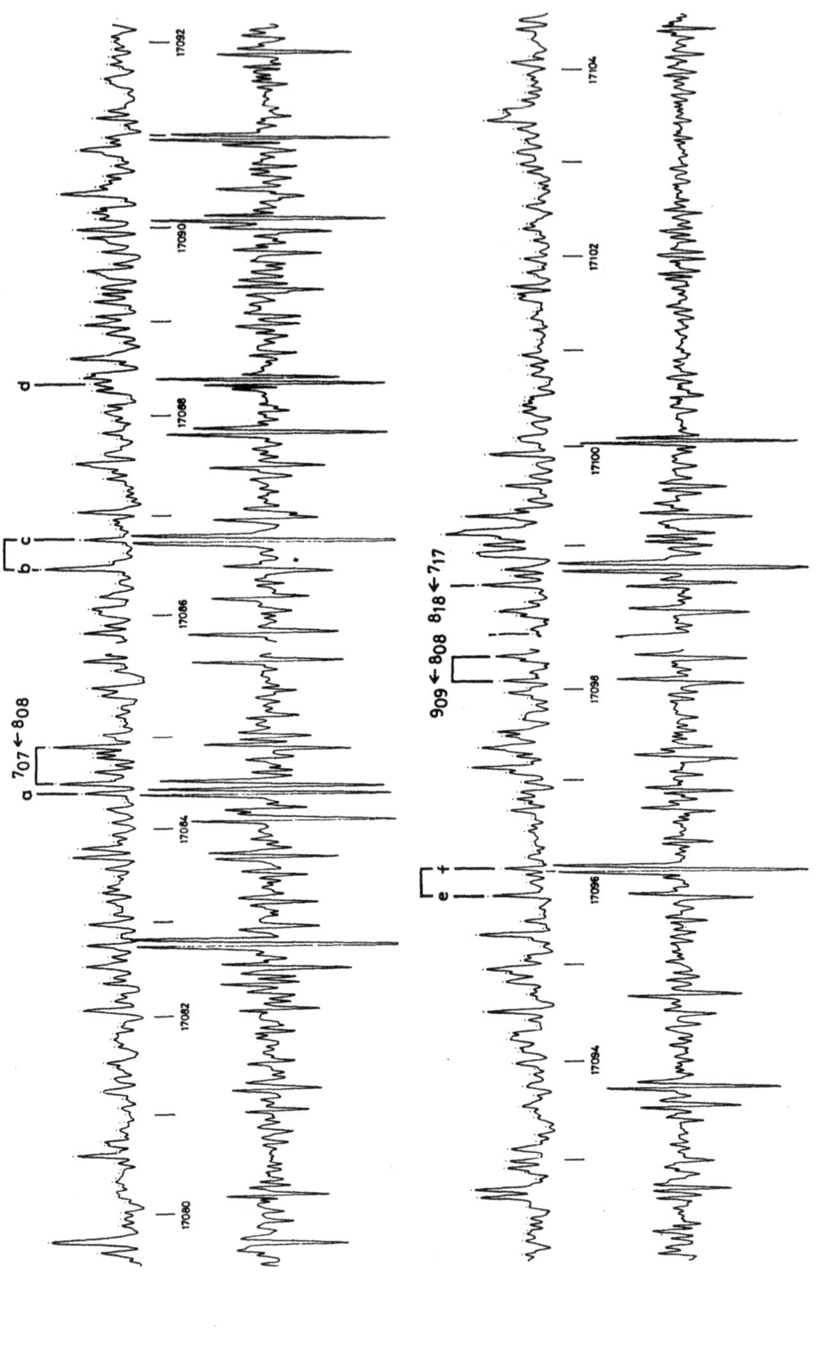

Fig. 2.6. The absorption spectrum (*upper trace*) and the Stark modulation spectrum (*lower trace*) of NO_2 around 585 nm. The frequency marks are given at 1 cm^{-1} intervals

It should be noted that the sign of the Stark shift in all the $7_{07} \leftarrow 8_{08}$ and $9_{09} \leftarrow 8_{08}$ transitions is positive. (The sign can be determined by comparing the line shape with the first-order Stark pattern.) This is explained by the downward Stark shift of the 8_{08} level in the ground vibronic state. On the contrary, the negative signs of the $8_{18} \leftarrow 7_{17}$ and $6_{16} \leftarrow 7_{17}$ transitions are due to the upward shift of the 7_{17} level. These observed Stark shifts in the 593.3-nm band confirm the earlier assignments [2.15–17].

In Fig. 2.6 two pairs of lines b, c and e, f have similar line shapes and separations, suggesting that they share a common upper state. The two pairs are separated by $9.28\,\mathrm{cm}^{-1}$, which agrees with the calculated 6_{06}–4_{04} separation in the ground vibronic state. Thus $5_{05} \leftarrow 6_{06}$ and $5_{05} \leftarrow 4_{04}$ are assigned to these transitions.

The first-order Stark pattern of lines c and f indicates that the upper spin component of the 5_{05} state is accidentally nearly degenerate with some other state which perturbs the 5_{05} state through the Stark effect. We tentatively assign this perturbing level to one spin component of the 4_{14} state in the same vibronic state. If this assignment is correct, then transitions terminating on the 4_{14} state, such as the $4_{14} \leftarrow 5_{15}$ and $4_{14} \leftarrow 4_{13}$ transitions in the same band, should also show the first-order Stark pattern. Lines a and d are identified as these transitions, because the a–d separation agrees with the calculated 5_{15}–4_{13} separation in the ground state. Moreover, the a–c separation which agrees with the 5_{15}–6_{06} separation in the ground state supports this assignment. Further analysis of the Stark modulation spectrum will be published elsewhere [2.19].

Recently, quantitative measurements of the Stark shift for some lines in the 593.3-nm band were made in a level-crossing experiment by *Bylicki* and *Weber* [2.20].

3. Progress in Spectroscopic Studies of NO_2 Using Lasers

In this chapter, recent spectroscopic studies of NO_2 using lasers are briefly reviewed. Details of various techniques of laser spectroscopy are described, for example, in a book by *Demtröder* [3.1]. Current topics of laser spectroscopy are found in the Proceedings of the International Conferences on Laser Spectroscopy [3.2,3]. A selected bibliography of recent studies, including non-laser experiments and theoretical work, is given in Chap. 4.

3.1 Laser-Induced Fluorescence

The fluorescence spectrum is expected to become very simple when one rovibronic level is selectively populated by narrow-band pumping. The analysis of the NO_2 visible spectrum has largely relied upon the laser-induced fluorescence spectrum.

The fluorescence spectrum of NO_2 excited by argon- and krypton-ion lasers, was first studied by *Sakurai* and *Broida* [3.4]. Subsequently, the fluorescence spectrum was excited with various wavelengths of argon-ion lasers [3.5–10], krypton-ion lasers [3.11], a helium-neon laser [3.7], a helium-cadmium laser [3.12], and dye lasers [3.13–17].

Under medium resolution, the fluorescence spectrum reflects both the vibrational and rotational structures of the ground electronic state. With a knowledge of the ground state molecular constants, the rotational quantum numbers N' and N'' in the optical transition can be determined from the frequency separation between the P and R branch lines in the fluorescence. The energy of the excited state level thus assigned is simply the sum of the initial state energy and the laser excitation energy. The rotational structure of the excited electronic state, therefore, can be determined from the fluorescence spectra excited by various laser lines. In the early extensive work along this line, the 593.3-nm region was covered by *Stevens* and *Zare* [3.15], the 612.5-nm region by *Lee* and *Peck* [3.17], and the 488- and 514.5-nm regions by *Schmiedl* et al. [3.10]. The vibrational structure of the ground electronic state was also studied by laser-induced fluorescence [3.7–9,11]. In [3.10], *Schmiedl* et al. used a molecular beam of NO_2 to reduce the Doppler width and thus to excite the single *hfs* levels in the upper state selectively.

Subsequently, rotational analyses of the 648- and 612.5-nm regions of the NO_2 absorption spectrum were performed by *Brand* et al. [3.18] and by *Monts* et al. [3.19], respectively, using laser-induced fluorescence. The vibrational structure of the ground electronic state of $^{15}NO_2$ has been investigated by *Brand* et al. [3.20] using several argon-ion-laser lines.

Laser-induced fluorescence from the $^2B_2$ state of NO_2, when exicted with high energy, shows anomalous features indicating coupling between the $^2B_2$ state and high vibrational levels of the ground electronic state [3.10,21]. In addition to vibronic coupling, which is the leading perturbation term, spin-orbit coupling has been considered as one of the possible coupling mechanisms between the two electronic states [3.22]. Definite evidence of spin-orbit coupling was found by *Brand* and his co-workers [3.23,24], who observed some transitions with the selection rule $\Delta N = 2$ in absorption or emission when the fluorescence was excited with wavelengths near 500 nm. They also observed anomalous intensity distributions in the P and R lines in some fluorescence bands [3.24]. This was interpreted by them as an indication that the intermediate state in the absorption-fluorescence cycle has both $^2B_2$ and $^2B_1$ basis functions in its total wavefunction.

The supersonic expansion technique was employed by *Raitt* et al. [3.25] to cool the molecules to $\cong$3 K and thus simplify the fluorescence spectrum of NO_2 excited by a tunable dye laser in the region 445–460 nm. In all cases studied, they observed fluorescence to the (0 0 1) and (0 1 1) vibrational levels of the ground electronic state, which is not expected for the C_{2v} symmetry assumed in the excited state.

Another consequence of coupling between the electronic states of NO_2 is the anomalously long radiative lifetime. Many fluorescence measurements yielded radiative lifetimes ranging from 15 to 100 μs, which are at least two orders of magnitude longer than the lifetime calculated from the integrated absorption coefficient [3.26]. On the other hand, short lifetimes of a few μs or less have been observed for excitation at certain wavelengths [3.27–29]. The wide range of observed lifetimes suggests that the degree of coupling between the electronic states varies significantly from one level to another.

Donnelly and *Kaufman* [3.30] measured the excitation energy dependence of the radiative lifetime, using pulsed Nd-YAG-laser excitation between 473 and 659 nm. The lifetime measured at low pressures changed smoothly from 88 μs at 473 nm to 120 μs at 562 nm. The fluorescence excited at longer wavelengths often showed nonexponential decays like those reported in earlier work. These observations were consistently explained by them in terms of the $^2B_2 - {}^2A_1$ vibronic coupling mechanism.

Chen et al. [3.31] used a supersonic molecular beam of NO_2 and a tunable pulsed dye laser to measure the lifetimes of various vibronic bands in the region 575–600 nm. Single exponential decay with a lifetime between 15 and 40 μs was observed for each excitation wavelength.

Imasaka et al. [3.32], and later *Donnelly* et al. [3.33], observed short-lived discrete vibrational bands and a long-lived continuum in the fluorescence spectrum of NO_2. The fast and slow processes were interpreted by *Donnelly* et al. as collision-induced rotational energy transfer and a subsequent stepwise vibrational relaxation in the excited state, respectively. This model was consistent with the temperature dependence of the fluorescence spectrum observed by *Keil* et al. [3.34]. More recently, *Takezawa* et al. [3.35,36] employed this model to explain the time-resolved, dispersed fluorescence spectrum excited in the blue region.

Often, the excitation spectrum (Sect. 3.2) and the fluorescence spectrum are taken together in one study because the experimental setup is similar for both measurements and they give information complementary to each other.

3.2 Laser Excitation Spectroscopy

An excitation spectrum is obtained by scanning the laser frequency while monitoring the fluorescence intensity. When the total fluorescence is monitored, the excitation spectrum usually agrees well with the corresponding absorption spectrum measured in transmission. (This agreement is seen, for example, between Chart No. 21 in this book and Fig. 1 of the paper by *Foth* et al. [3.37].) However, it is usually easier to achieve higher sensitivities in excitation spectra than in transmission spectra because of the absence of background light in fluorescence detection. Besides, use of the excitation spectrum is more feasible for combination with the molecular beam techniques than use of the transmission spectrum. When the fluorescence is monitored through a narrow bandwidth to select some particular band, the spectrum generally becomes simpler. The resulting "selective excitation spectrum" makes line assignment much easier.

High resolution excitation spectroscopy of NO_2 using a molecular beam was demonstrated by *Demtröder* et al. [3.38] in the 514.5-nm region. The residual Doppler width was 20 MHz and the hyperfine structure was resolved when a collimated molecular beam of NO_2 was crossed at right angles with a light beam from an argon-ion laser tunable within a 7-GHz range. In a study which extended this work, *Schmiedl* et al. [3.10] used both the 488- and 514.5-nm argon-ion-laser lines and identified many transitions in fluorescence from a combined analysis of the excitation spectra and laser-induced fluorescence spectra.

A drastic simplification of the NO_2 excitation spectrum was achieved by *Smalley* and co-workers [3.39–41] through the use of a supersonic molecular beam. The NO_2 sample seeded in a carrier gas of argon or helium was made to expand through a nozzle and cooled down, respectively, to $\cong$0.5 K rotationally and 150 K vibrationally. Using a tunable dye laser with a bandwidth of 1 GHz, they observed and analyzed 140 vibronic bands in the region 570.8–670.8 nm. Only a small number of transitions with low rotational quantum numbers appeared in each band. The major part of the excitation spectrum thus observed is reproduced in the book of *Hsu* et al.

[3.42]. The number, 140, of vibronic bands is much greater than can be expected for the calculated state density in the $^2B_2$ state. This discrepancy was attributed to vibronic coupling between the $^2B_2$ and $^2A_1$ basis states.

Recently, *Foth* et al. [3.37] performed very high resolution and high accuracy excitation spectroscopy of NO_2 in the 592.5-nm region (the region of the band No. 101 of *Smalley* et al.[3.40]) by combining the supersonic beam technique and a tunable dye laser with a linewidth down to 10 MHz. The absolute wavelengths of some selected lines were determined with a lambdameter to an accuracy of 10^{-3} cm^{-1}, while the separations of closely spaced lines were measured from wavelength marks obtained from a Fabry-Perot interferometer. The unambiguous rotational assignment was again possible through the fluorescence spectrum, which was monitored with a very high sensitivity by means of photon counting and signal averaging. Thorium emission lines were superimposed on the NO_2 fluorescence spectrum as wavelength references. The high accuracy in wavelength measurement allowed a precise analysis of the rotational, fine, and hyperfine structures in the $^2B_2$ electronic state. Particularly, it was found that the fine structure parameter ε' is not constant but changes with the rotational quantum number N', as expressed well by $\varepsilon' = \alpha N' + \beta$. On the other hand, the hyperfine structure was fit satisfactorily with a common value of the Fermi-interaction constant about half the value of the ground-state constant.

In an extended study of this high resolution work, *Vedder* et al. [3.43] assigned about 170 fine structure components of rotational transitions in four vibronic bands (bands No. 101–104 in [3.40]) in the 591.4- to 592.9-nm region. These studies, combined with an earlier study of the 593.3-nm band by *Stevens* and *Zare* [3.15], provide a nearly complete analysis of one strong absorption feature in the red region of the NO_2 spectrum. *Foth* and *Vedder* [3.44] observed spin forbidden lines in the excitation spectrum in the 592.5-nm region. The observed line intensities for these transitions could be fit by a superposition of the J- and G-coupling schemes.

Time-resolved excitation spectroscopy techniques have been developed by *Imasaka* et al. [3.45]. In earlier work [3.32] they investigated the time-resolved fluorescence of NO_2 excited by pulsed dye lasers and found that the appearance of the fluorescence spectrum changes with time because of the difference of lifetimes between the discrete lines and the continuum. With an appropriate selection of the gate time and the spectral range in which the fluorescence was monitored, they were able to obtain a discrete and simplified structure in the excitation spectrum of NO_2 in the 455-nm range. Later, the same technique was employed by *Takezawa* et al. [3.46–48] to analyze the rotational structure of several bands between 435 and 500 nm.

Chen et al. [3.49] used laser excitation spectra to determine the dissociation limit of rotationally cooled NO_2. The onset of predissociation was located at 397.82 ± 0.02 nm, as compared with a previous value of 397.9 ± 0.1 nm obtained by *Douglas* and *Huber* [3.50].

3.3 Raman Spectroscopy

Intense, monochromatic radiation from lasers makes observation of Raman spectra easier. So far, however, a rather small number of works on the Raman spectrum of NO_2 has been reported, probably because the Raman spectrum is intermingled with intense fluorescence lines and because it is therefore difficult to extract new information from the Raman spectrum.

Marsden and *Bird* [3.51] investigated the resonance Raman scattering of NO_2 with an argon-ion laser and succeeded in quenching the fluorescence lines considerably by using high pressure samples. Then the rotational Raman lines were separated from the remaining fluorescence lines by comparing the Raman spectra for different excitation wavelengths. From the selection rules operating in the observed rotational Raman spectra, *Bird* and *Marsden* [3.52] determined definitely the excited electronic state symmetries, $^2B_1$ and $^2B_2$, of NO_2 excited at 488 and 514.5 nm.

Bist et al. [3.53] observed in the resonance Raman spectrum of NO_2, pressured to 1 atm with Ne, many combinations and overtones of the three fundamental vibrations in the ground electronic state.

Tevault and *Andrew* [3.54] examined resonance Raman scattering of matrix-isolated NO_2. The fluorescence was quenched and they could analyze the vibrational structure of the resonance Raman spectrum. Recently, *Bolduan* and *Jodl* [3.55] investigated resonance Raman scattering of NO_2 in Ar, Kr, and Xe matrices. They determined the vibration energies of the ground electronic state of NO_2 in each matrix and obtained a nearly complete set of vibrational constants. Raman spectra of matrix-isolated NO_2 at high pressures up to 8 kbar were studied by *Jodl* and *Holzapfel* [3.56].

Coherent anti-Stokes Raman emission has been also observed [3.57,58].

3.4 Double Resonance and Level-Crossing Spectroscopy

In double resonance experiments the changes in one transition caused by radiation resonant to another transition are detected. The detection is made through absorption, emission, or fluorescence. Each of the two transitions can be, in principle, in any spectral region. The intense coherent light of lasers available at various wavelengths has therefore made a great variety of double resonance experiments possible. A good review of the theoretical and experimental work on double resonance laser spectroscopy of molecules was given by *Shimoda* [3.59].

Microwave-optical double resonance spectroscopy has been a powerful technique in the identification of the complicated visible spectrum of NO_2. The rotational structures in both the excited and ground electronic states have been determined by this technique [3.6,14,60–64].

This microwave-optical double resonance technique was further employed by *Solarz* and *Levy* [3.65] to study the dc Stark and Zeeman effects in the hyperfine transitions in the $^2B_2$ electronic state and microwave saturation behavior of the double resonance signals. Infrared-optical double resonance in NO_2 was observed for the first time by *Herman* et al. [3.66]. They used an argon-ion laser to excite molecules to the $^2B_2$ and $^2B_1$ states. An excited vibronic state transition induced by a CO laser was then detected by monitoring the visible fluorescence. Their observation suggested that the isoenergetic collisional transfer contributes to the double resonance and that the excited levels participating in the double resonance have very short lifetimes of 0.6 ± 0.2 μs. Recently *Khoobehi* and *Roberts* [3.67] used infrared-microwave double resonance to measure the temporal population changes in selected rotational levels in the (0 0 0) and (0 1 0) vibrational states of NO_2 under continuous exposure to the CO_2-laser output.

Allegrini et al. [3.68] applied infrared-radiofrequency double resonance to the detection of the magnetic-dipole spin-rotation transitions within the ground and $v_3 = 1$ excited vibrational states of NO_2. They placed an absorption cell, constructed as an rf transmission line, inside the optical cavity of a CO laser and detected rf transmission through the changes in the laser output power. *Brown* et al. [3.69] observed similar magnetic-dipole transitions in the ground vibrational state by the microwave-optical double resonance technique. They detected the microwave transitions by monitoring the fluorescence induced by the argon-ion-laser output.

A series of double resonance and level-crossing experiment have been done by *Weber* and his co-workers [3.70–77] to study an unresolved problem in NO_2, i.e., the "Hanle effect anomaly." As reported by *Figger* et al. [3.78] and *Bonilla* and *Demtröder* [3.79], the $g\tau_c$ product, where g is the magnetic moment of the excited state and τ_c its dephasing time, is about ten times smaller than that expected from the lifetime τ_R of fluorescence decay. *Weber* et al. [3.70] measured the g values for some levels of the 612.6- and 593.3-nm bands using optical-radiofrequency double resonance. The separation between the magnetic sublevels at a known magnetic field strength was measured. It was found that the g values behave as predicted for the Hund coupling case b and hence are not the cause of the Hanle effect anomaly.

To explain this discrepancy between lifetimes determined from the two kinds of experiments, *Weber* et al. [3.80] proposed a model of the NO_2 level structure. The model assumed that each upper level in the optical transitions has a finer structure called a "bunch structure", consisting of sublevels whose spacing is about 10 MHz or less. They showed with this model that an intramolecular dephasing process causes the coherent time in the Hanle measurements to differ from the radiative lifetime. In addition, this model predicted that the radiation excitation prepares NO_2 in a nonstationary state with a lifetime of 0.5 μs. A formulation of this nonstationary state was given by *Weber* [3.81] on the assumption of the presence of inversion splitting caused by the asymmetric structure of NO_2, i.e., two unequal N-O bond lengths in the excited state as suggested by several authors [3.62,82,83].

The lifetime discrepancy was finally removed by *Bylicki* and *Weber* [3.73] when they observed in a molecular beam Hanle experiment very narrow linewidths, which imply that $\tau_c = \tau_R$. In a similar experiment with a static gas sample they found that the linewidth increases very quickly with increasing pressure until it gradually goes onto a less steep straight line at pressures higher than 0.5μbar. They concluded that the broad linewidth obtained in an earlier experiment [3.78] resulted from an invalid extrapolation to zero pressure of the linewidths measured at higher pressures. However, the narrow linewidths they observed are still contradictory to the result of another molecular beam experiment done by *Bonilla* and *Demtröder* [3.79].

In subsequent optical-radiofrequency double resonance and level-crossing experiments using a molecular beam, *Bylicki* and *Weber* [3.74,77] found that the g factor of the $^2B_2$ state of NO_2 depends on the quantum number K_a of the near-prolate rotor. The deviation from the Hund coupling case b was attributed to perturbations by the spin-orbit interaction.

The coherence properties of NO_2 have been investigated recently by *Brucat* and *Zare* [3.84], who looked at its Zeeman quantum beats. Their observations showed that the excited state population and coherence decay at equal rates, in accordance with the above results obtained by *Bylicki* and *Weber* [3.73]. However, no evidence for magnetic field decoupling as observed in [3.73,74] was found. In the same experiment, *Brucat* and *Zare* [3.84] observed an irregular deviation of the excited state Landé g factor from the Hund coupling case b.

3.5 Diode-Laser and Color-Center-Laser Applications

Semiconductor-diode lasers are useful tunable sources for high resolution infrared spectroscopy. The diode lasers cover a wide spectral range, 0.6–30 μm, although each diode laser is tunable only in a narrow interval of a few tens of wave numbers. A temperature change produced by external cooling is most commonly used for wavelength tuning. An electric current change is a convenient method of fine tuning, although it causes the laser output power to change at the same time. With appropriate frequency stabilization, spectral resolution down to the kilohertz range has been achieved.

Flaud et al. [3.85] observed about 380 lines in the ν_2 band of NO_2, including the completely resolved Q branches, with a high resolution of 0.002 cm^{-1}, by using the diode-laser spectrometer developed by *Reddy* et al. [3.86]. An improved set of rotational and spin-rotation constants for the (0 1 0) vibrational state was obtained from the measurement. The high resolution spectrum of the ν_2 band enabled *Devi* et al. [3.87] to determine absolute line intensities, transition moments, and broadening parameters. Similar measurements on the ν_3 and $\nu_2 + \nu_3 - \nu_2$ bands made it possible to determine the transition moments for these bands and the broadening parameters for the ν_3 band as well as the spin-rotation constants for the (0 0 1) and (0 1 1) states [3.88,89]. These line parameters are important for a precise determination of the NO_2 concentration in the atmosphere and combustion products.

Reid et al. [3.90] used a diode laser and a multipass White cell to detect NO_2 in an atmosphere with a sensitivity of better than 100 ppt.

The color-center laser is another laser which has a wide tunable range and a high spectral purity. The range 2.3–3.3 μm can be covered by using color centers in Na-doped KCl and other materials. *Dillenschneider* and *Curl* [3.91] investigated the magnetic rotation spectrum of the $\nu_1 + \nu_2 + \nu_3$ band of NO_2 using a computer-controlled color-center laser, which is capable of a wide scan exceeding 10 cm^{-1} with a 3-MHz linewidth. The spin splitting in this band was resolved for the first time and molecular constants were determined for the (1 1 1) vibrational level.

3.6 Laser Magnetic Resonance

Laser magnetic resonance (LMR) spectroscopy is an extension of electron paramagnetic resonance spectroscopy to higher frequency regions by using a laser in place of a microwave or radiowave source. In laser magnetic resonance, absorption is observed when the Zeeman components of some transition are brought to the laser frequency by changing the magnetic field applied to the sample. Laser magnetic resonance spectroscopy using many fixed-frequency infrared lasers has been very powerful, although only molecules with an unpaired electron spin can be studied by this method. From LMR one can obtain information about magnetic properties as well as accurate molecular constants of molecules and radicals, complementary to those obtainable in conventional infrared grating spectroscopy. LMR was first applied to an observation of magnetic resonance in ground state O_2 at the 337-μm HCN-laser frequency [3.92].

The first application of LMR to the NO_2 molecule was made by *Curl* et al. [3.93]. They observed several rotational transitions in the ground vibrational state of NO_2 using the 311- and 337-μm HCN-laser lines. *Freund* et al. [3.94] used CO-laser lines to study vibration-rotation transitions in the ν_3 band of $^{14}NO_2$ and $^{15}NO_2$. Recently, *Hakuta* and *Uehara* [3.95] observed magnetic resonance spectra of the ν_2 band of NO_2 using a CO_2 laser. Many Zeeman components observed in 19 vibration-rotation transitions made it possible to determine accurate molecular parameters including the electron g factors for the $v_2 = 1$ excited vibrational state.

3.7 Other Studies

Intracavity laser spectroscopy is known as a very sensitive method for detecting absorption. *Baev* et al. [3.96] used this method to investigate the absorption spectrum of the atmosphere of the laboratory room. With a cw dye laser and a spectrograph, they detected more than 700 absorption lines in the 583- to 605-nm region including many NO_2 lines.

Optoacoustic spectroscopy is also capable of sensitive detection of absorption in a gas sample. *Angus* et al. [3.97] and *Ioli* et al. [3.98] have achieved very high sensitivities to detect low-concentration NO_2 in air down to 5 ppb.

Polarization spectroscopy [3.99] is a very powerful method used to simplify complicated molecular spectra. In this method, one transition is selectively pumped by a narrow-band polarized pulse, causing uneven populations among the magnetic sublevels of the initial and final states. This population gradient gives rise to birefringence and dichroism at wavelengths of all transitions to and from these selected states. As a result, a linearly polarized probe beam passing through the sample can reach the detector through a crossed polarizer at those specified wavelengths.

Teets et al. [3.100] applied polarization spectroscopy to NO_2 and demonstrated that the vibrational and rotational structures in the excited state can be readily determined from the simplified spectrum. Unlike the case of rotationally cooled spectra [3.40], observations are not limited to low rotational quantum numbers.

Later, polarization spectroscopy was employed by *Brand* and co-workers [3.101–103] to analyze the rotational structure of many bands in the ranges 16600–17000 and 17350–17850 cm^{-1}. At the same time, they observed stimulated emission from some of the pumped excited states [3.102].

Optogalvanic spectroscopy is based on the effect whereby the impedance of atomic or molecular species in discharges or flames is affected by irradiation of light resonant to certain transitions in the species. The visible optogalvanic spectrum of NO_2 has been studied by *Feldman* [3.104] and *Suzuki* [3.105] using dc and rf discharges, respectively, and has been found to show essentially the same structures as the absorption spectrum. Recently, *Webster* and *Menzies* [3.106] observed the infrared optogalvanic spectrum of the NO_2 ν_3 band at 6.2 μm using a diode laser.

A number of studies of multiphoton excitation, dissociation, and ionization spectroscopy have been performed using various kinds of laser. Recent progress in this field has been reported in [3.107–116] and in references therein.

4. Selected Bibliography 1978–1983

This chapter presents a list of literature of spectroscopic studies on the NO_2 molecule which were published between 1978 and the beginning of 1984. The list has about 130 selected papers, including those reporting non-laser experiments and theoretical work which were not referred to in Chap. 3. All papers are given with full titles. The papers were searched for mainly through the CAS Database. A compilation of earlier work before 1978 was given by *Hsu* et al. [4.1].

Papers are sorted into seven groups:

- Theoretical studies and data compilation
- Microwave and millimeter-wave spectroscopy
- Infrared spectroscopy
- Visible and uv spectroscopy ($\lambda < 1\ \mu m$)
- Double resonance and level-crossing spectroscopy
- Fluorescence lifetime and cross-section measurements
- Other experimental studies

All theoretical studies are listed in Sect. 4.1. In each group, papers are arranged in alphabetical order by the first author's name.

4.1 Theoretical Studies and Data Compilation

Baird, N.C., Taylor, K.F.: *Ab initio* MO predictions for the geometries and energies of the Σ and Π states of some conjugated free radicals. Can. J. Chem. **58**, 733 (1980)

Baird, N.C., Taylor, K.F.: The stabilizing effect of *d* orbitals on the central nitrogen atom in nitrogen-oxygen molecules and ions. Chem. Phys. Lett. **80**, 83 (1981)

Benioff, P.A.: *Ab initio* calculations on NO_2 and NO_2^-: optimization of diffuse Gaussian exponents. Theor. Chim. Acta **48**, 337 (1978)

Benioff, P.A.: *Ab initio* calculations of the vertical electronic spectra of NO_2, NO_2^+, and NO_2^- . J. Chem. Phys. **68**, 3405 (1978)

Bermejo, M.R.: Possible explanation of the structure of some polyatomic inorganic molecules. Acta Cient. Compostelana **16**, 183 (1979)

Clary, D.C.: Potential energy functions of polyatomic molecules. Chem. Phys. **41**, 387 (1979)

Efremov, Yu.S., Zhirnov, N.I.: Calculation of the spectrum of valence vibrations of nonlinear triatomic molecules in an approximation of partial

Morse oscillators. Opt. Spektrosk. **46**, 401 (1979) [English transl.: Opt. Spectrosc. **46**, 222 (1979)]

Gerloff, M., Russegger, P., Brickmann, J.: On the coupling of vibronic states in NO_2. Ber. Bunsenges. Phys. Chem. **83**, 108 (1979)

Handy, N. C., Goddard, J. D., Schaefer,H. F., III: Generalization of the direct configuration interaction method to the Hartree-Fock interacting space for doublets, quartets, and open-shell singlets. Applications to NO_2 and NO_2^-. J. Chem. Phys. **71**, 426 (1979)

Hoy, A.R.: High order coriolis interactions in NO_2. J. Mol. Spectrosc. **86**, 55 (1981)

Innes, K.K.: Vibronic coupling as a major cause of the anharmonicity of antisymmetric stretching in the ground state of NO_2. J. Mol. Spectrosc. **96**, 331 (1982)

Jackels, C.F.: Theoretical excitation energies of the lowest quartet states of nitrogen dioxide. J. Chem. Phys. **70**, 4664 (1979)

Lattanzi, F.: The Zeeman effect in molecular doublet-doublet transitions in the "high field-low field" limit. J. Phys. Lett. **43**, 383 (1982)

Lyaptsev, A.V., Petrun'kin, A.M., Kiselev, A.A.: Manifestation of spin-vibration-rotation interactions in molecular spectra. Opt. Spektrosk. **49**, 493 (1980) [English transl.: Opt. Spectrosc. **49**, 268 (1980)]

Morita, K., Sato, H., Ishiguro, E.: *Ab initio* LCAO MO SCF calculation of potential surface of the NO_2 molecule. Int. J. Quantum Chem. **18**, 151 (1980)

Pavlyuchko, A. I., Lozenko, G. F., Gribov, L. A.: Interpretation of states and analysis of changes in the geometry of molecules during anharmonic vibrational excitations. Opt. Spektrosk. **50**, 450 (1981) [English transl.: Opt. Spectrosc. **50**, 243 (1981)]

Pavlyuchko, A. I., Lozenko, G. F., Khovrin, G. V., Gribov, L. A.: Solution to the anharmonic vibration problem for H_2O, H_2Se, NO_2, and O_3 molecules by a variation method. Opt. Spektrosk. **52**, 64 (1982) [English transl.: Opt. Spectrosc. **52**, 37 (1982)]

Perrin, A., Flaud, J.-M., Camy-Peyret, C.: Calculated line positions and intensities for the $\nu_1 + \nu_3$ and $\nu_1 + \nu_2 + \nu_3 - \nu_2$ bands of $^{14}N^{16}O_2$. Infrared Phys. **22**, 343 (1982)

Rakshit, S.C., Banerjee, M.: Stable confirmations of nitrogen dioxide, nitrous oxide and nitrosyl fluoride. Indian J.Chem., Sect. A **17A**, 329 (1979)

Rothman, L.S., Gamache, R.R., Barbe, A., Goldman, A., Gillis, J.R., Brown, L..R., Toth, R.A., Flaud, J.-M., Camy-Peyret, C.: AFGL atmospheric absorption line parameters compilation: 1982 edition. Appl. Opt. **22**, 2247 (1983)

Rothman, L.S., Goldman, A., Gillis, J.R., Tipping, R.H., Brown, L.R., Margolis, J.S., Maki, A.G., Young,L.D.G.: AFGL trace gas compilation: 1980 version. Appl. Opt. **20**, 1323 (1981)

Schirmer, J., Cederbaum, L.S., von Niessen, W.: Two-hole-one-particle configuration interaction approach for the ionization of open-shell molecules: application to NO_2. Chem. Phys. **56**, 285 (1981)

Stannard, P.R., Elert, M.L., Gelbart, W.M.: On the overtone-combination spectra of XY_2 molecules. J. Chem. Phys. **74**, 6050 (1981)

Suzuki, I.: Anharmonic potential functions of simple molecules. III. Computation of vibration-rotation energies of XYZ and X_2Y_2 type linear molecules through direct numerical diagonalization: application to the NO_2 molecule. Bull. Chem. Soc. Jpn. **52**, 1606 (1979)

Voitsekhovskaya, O.K., Cherepanov, V.N.: Intensities of fine structure lines of a rotational spectrum of nitrogen dioxide. Opt. Spektrosk. **50**, 280 (1981) [English transl.: Opt. Spectrosc. **50** 148 (1981)]

Weber, H.G.: Nonstationary states in molecular spectroscopy: the Hanle effect in NO_2. Z. Phys. A **293**, 203 (1979)

Weber, H.G., Brucat, P.J., Zare, R.N.: A simple model for the difference between coherence time and radiative lifetime in NO_2. Chem. Phys. Lett. **60**, 179 (1979)

4.2 Microwave and Millimeter-Wave Spectroscopy

Bowman, W.C., De Lucia, F.C.: The millimeter and submillimeter spectrum of NO_2: a study of electronic effects in a nonsinglet light asymmetric rotor. J. Chem. Phys. **77**, 92 (1982)

Meier, D.: Pressure broadening of the nitrogen dioxide hyperfine multiplet at 93 GHz. J. Quant. Spectrosc. Radiat. Transfer **20**, 129 (1978)

Moazzen-Ahmadi, M.N., Roberts, J.A.: Linewidth parameters in the rotational spectrum of nitrogen dioxide. J. Mol. Spectrosc. **96**, 336 (1982)

Morino, Y., Tanimoto, M., Saito, S., Hirota, E.: Microwave spectrum of nitrogen dioxide in excited vibrational states — equilibrium structure. J. Mol. Spectrosc. **98**, 331 (1983)

Nilges, M., Shiotani, M., Yu, C.T., Barkley, G., Kera, Y., Freed, J.H.: Ultrahigh vacuum ESR studies on clean metal surfaces: NO_2 on copper. J. Chem. Phys. **73**, 588 (1980)

Tanaka, N., Hamada, Y., Tsuboi, M., Saito, S., Endo, Y., Hirota, E.: Microwave spectrum of $^{15}N^{16}O_2$ in the ground state. J. Mol. Spectrosc. **103**, 87 (1984)

4.3 Infrared Spectroscopy

Dana, V., Maillard, J.P.: Analysis of the $\nu_1 + \nu_3$ band of $^{14}N^{16}O_2$. J. Mol. Spectrosc. **71**, 1 (1978)

Devi, V. Malathy, Das, Palash P., Bano, A., Rao, K. Narahari, Flaud, J.-M., Camy-Peyret, C., Chevillard, J.P.: Diode laser measurements of intensities, N_2-broadening, and self-broadening coefficients of lines of the ν_2 band of $^{14}N^{16}O_2$. J. Mol. Spectrosc. **88**, 251 (1981)

Devi, V. Malathy, Fridovich, B., Jones, G.D., Snyder, D.G.S., Neuendorffer, A.: Temperature dependence of the widths of N_2-broadened lines of the ν_3 band of $^{14}N^{16}O_2$. Appl. Opt. **21**, 1537 (1982)

Devi, V. Malathy, Fridovich, B., Jones, G.D., Snyder, D.G.S., Das, Palash P., Flaud, J.-M., Camy-Peyret, C., Rao, K. Narahari: Tunable diode laser spectroscopy of NO_2 at 6.2 μm. J. Mol. Spectrosc. **93**, 179 (1982)

Dillenschneider, W., Curl, R.F., Jr.: Color center laser spectroscopy of $\nu_1 + \nu_2 + \nu_3$ of NO_2 using magnetic rotation. J. Mol. Spectrosc. **99**, 87 (1983)

Flaud, J.-M., Camy-Peyret, C., Devi, V. Malathy, Das, Palash P., Rao, K. Narahari: Diode laser spectra of the ν_2 band of $^{14}N^{16}O_2$: the (0 1 0) state of NO_2. J. Mol. Spectrosc. **84**, 234 (1980)

Hakuta, K., Uehara, H.: Infrared laser magnetic resonance (LMR) spectroscopy of NO_2 (ν_2) with a CO_2 laser. J. Mol. Spectrosc. **94**, 126 (1982)

Holland, R.F., Maier, W.B.: Infrared absorption spectra of nitrogen oxides in liquid xenon. Isomerization of N_2O_3. J. Chem. Phys. **78**, 2928 (1983)

Laurin, M., Cabana, A.: High resolution infrared spectrum of ν_1 band of $^{14}N^{16}O_2$. J. Mol. Spectrosc. **69**, 421 (1978)

Lubezky, A., Folman, M.: Infrared spectra of NO, N_2O, NO_2 and O_2 adsorbed on SiO films. J. Chem. Soc., Faraday Trans. 1 **74**, 2935 (1978)

Perrin, A., Mandin, J.-Y., Camy-Peyret, C., Flaud, J.-M., Chevillard, J.-P.: The ν_1 band of $^{14}N^{16}O_2$: line positions and intensities. J. Mol. Spectrosc. **103**, 417 (1984)

Reid, J., El-Sherbiny, M., Garside, B.K., Ballik, E.A.: Sensitivity limits of a tunable diode laser spectrometer, with application to the detection of NO_2 at the 100-ppt level. Appl. Opt. **19**, 3349 (1980)

Sulzmann, K.G.P., Hamins, A., Tang, S.Y.: 3.508 μ-He-Xe laser line absorption by formaldehyde monomer and dimer and by nitrogen dioxide monomer and dimer without and in the presence of argon. J. Quant. Spectrosc. Radiat. Transfer **29**, 31 (1983)

Toth, R.A., Hunt, R.H.: Line strengths, spin-splittings, and forbidden transitions in the (1 0 1) band of $^{14}N^{16}O_2$. J. Mol. Spectrosc. **79**, 182 (1980)

Webster, C.R., Menzies, R.T.: Infrared laser optogalvanic spectroscopy of molecules. J. Chem. Phys. **78**, 2121 (1983)

4.4 Visible and uv Spectroscopy ($\lambda < 1\ \mu$m)

Armstrong, M.J., Brand, J.C.D., di Lauro, C.: Single-mode laser fluorescence of NO_2 excited at 488 nm. Can. J. Phys. **59**, 559 (1981)

Baev, V.M., Belikova, T.P., Ippolitov, M.B., Sviridenkov, E.A., Suchkov, A.F.: Absorption spectrum of the atmosphere in the 583-605 nm range obtained by the method of intracavity laser spectroscopy. Opt. Spektrosk. **45**, 58 (1978) [English transl.: Opt. Spectrosc. **45**, 31 (1978)]

Bolduan, F., Jodl, H.J.: Resonance Raman spectroscopy of matrix-isolated NO_2. J. Mol. Spectrosc. **91**, 404 (1982)

Brand, J.C.D., Chiu, P.H.: Spin-orbit-induced transitions in the laser-excited fluorescence of nitrogen dioxide. J. Mol. Spectrosc. **75**, 1 (1979)
Brand, J.C.D., Chiu, P.H., Hoy, A.R.: Laser fluorescence and high vibrational levels of $^{15}NO_2$. Can. J. Phys. **57**, 828 (1979)
Brand, J.C.D., Cross, K.J., Ernsting, N.P.: Polarization spectroscopy of NO_2. Chem. Phys. **59**, 405 (1981)
Brand, J.C.D., Cross, K.J., Ernsting, N.P., Yamashita, A.B.: Stimulated emission enhanced polarization labelling. Opt. Commun. **37**, 178 (1981)
Brand, J.C.D., Cross, K.J., Hoy, A.R.: Rotational analysis of the 6480 Å band absorption of NO_2. Can. J. Phys. **57**, 428 (1979)
Brand, J.C.D., Ernsting, N.P.: Rotational analysis of visible bands of nitrogen dioxide using polarization spectroscopy. J. Mol. Spectrosc. **91**, 389 (1982)
Bures, J., Leonard, F., Monchalin, J.-P.: Utilisation d'une ligne de microphotodiodes en spectroscopie dispersive: application à la détection du NO_2. Can. J. Phys. **61**, 301 (1983)
Chen, C.H., Clark, D.W., Payne, M.G., Kramer, S.D.: Determination of dissociative limit of NO_2 and characteristics of a molecular beam by fluorescence excitation spectra. Opt. Commun. **32**, 391 (1980)
Feldmann, D.: Opto-galvanic spectroscopy of some molecules in discharges: NH_2, NO_2, H_2, and N_2. Opt. Commun. **29**, 67(1979)
Foth, H.-J., Vedder, H.J.: Spin forbidden transitions in the $K_a = 0$ subband of NO_2 in the $\lambda = 592.5$ nm region. J. Mol. Spectrosc. **102**, 148 (1983)
Foth, H.-J., Vedder, H.J., Demtröder, W.: Sub-Doppler laser spectroscopy of NO_2 in the $\lambda = 592.5$ nm region. J. Mol. Spectrosc. **88**, 109 (1981)
Guthals, D.M., Gross, K.P., Nibler, J.W.: Resonant CARS spectra of NO_2. J. Chem. Phys. **70**, 2393 (1979)
Hardwich, J.L.: Collisional transfer of rotational energy in the $^2B_1$ electronic state of nitrogen dioxide. J. Mol. Spectrosc. **106**, 186 (1984)
Imasaka, T., Ogawa, T., Ishibashi, N.: Time-resolved spectroscopy on the excited electronic states of NO_2 in the neighborhood of 454.6 nm. J. Chem. Phys. **70**, 881 (1979)
Ioli, N., Violino, P., Meucci, M.: A simple transversely excited spectrophone. J. Phys. E **12**, 168 (1979)
Jodl, H.J., Holzapfel, W.B.: Raman spectra of NO_2^- in KI and of NO_2 in Ar at high pressure and at low temperatures. Chem. Phys. Lett. **55**, 259 (1978)
Koizumi, H., Hadeishi, T., Mclaughlin, R.D.: Detection of small molecules by magnetically tuned frequency-modulated atomic line sources. Appl. Phys. Lett. **34**, 382 (1979)
Kuwabara, S., Kuwata, K., Nishiyama, I., Hanazaki, I.: Acoustically oscillating emission from NO_2^* produced by infrared photosensitized reaction in SF_6+NO_2. Chem. Phys. Lett. **106**, 540 (1984)

Lakshman, S.V.J., Rao, J. Lakshmana: Vibrational modes of hydrogen and nitrogen dioxide molecules in the optical spectrum of nickel 1,10-phenanthroline crystal. Spectrosc. Lett. **11**, 637 (1978)

McIlwain, M.E., Hindman,J.C.: Discrete resonance CARS emission from NO_2. J. Chem. Phys. **73**, 68 (1980)

Merer, A.J., Hallin, K.E.J.:The temperature dependence of the 8920 Å band of NO_2 and the location of the 000–000 band of the $\tilde{A}^2B_2 - \tilde{X}^2A_1$ transition. Can. J. Phys. **56**, 838 (1978)

Merer, A.J., Hallin, K.E.J.: Rotational structure in the NO_2 absorption spectrum between 8170 and 8280 Å . Can. J. Phys. **56**, 1502 (1978)

Monts, D.L., Soep, B., Zare, R.N.: Rotational analysis of the NO_2 6125 Å region. J. Mol. Spectrosc. **77**, 402 (1979)

Morioka, Y., Masuko, H., Nakamura, M., Sasanuma, M., Ishiguro, E.: Absorption spectrum and absorption coefficients of NO_2 in the vacuum ultraviolet region. Can. J. Phys. **56**, 962 (1978)

Perrin, A., Camy-Peyret, C., Flaud, J.-M., Luc, P.: Rotational analysis of the 7390 and 7937 Å bands of NO_2 by means of Fourier transform spectroscopy. J. Mol. Spectrosc. **88**, 237 (1981)

Perrin, A., Flaud, J.-M., Camy-Peyret. C., Luc, P.: Rotational analysis of the 13200 and 13400 cm^{-1} bands of NO_2 by means of Fourier transform spectroscopy. Can. J. Phys. **60**, 1288 (1982)

Pommereau, J.-P., Hauchecorne, A.: Observations spectroscopique depuis le sol du dioxyde d'azote atmosphérique. C. R. Hebd. Seances Acad. Sci., Ser. B **288**, 135 (1979)

Raitt, I., Griffiths, A.M., Freedman, P.A.: Resonance fluorescence from nitrogen dioxide cooled in a supersonic jet. Chem. Phys. Lett. **77**, 433 (1981)

Saito, K., Higashihara, T., Nakamura, Y., Murakami, I.: Measurements of light emission from shock-heated CS_2, SO_2, and NO_2. J. Sci. Hiroshima Univ., Ser. A: Phys. Chem. **43**, 119 (1979)

Sugimoto, N.: Intensity distributions of fluorescence from the $^2B_1$ state of NO_2 excited at 495.0, 474.0, 454.5, and 436.7 nm. J. Mol. Spectrosc. **106**, 362 (1984)

Sugimoto, N., Takeuchi, N., Iijima, I., Arai, T., Takezawa, S.: Observation of spin splittings in the $^2B_1$ state of NO_2 by means of polarization spectroscopy. Chem. Phys. Lett. **106**, 403 (1984)

Sugimoto, N., Takeuchi, N., Takezawa, S.: Rotational analysis of the $^2B_1$ $(K' = 1) - {}^2A_1(K'' = 0)$ sub-band of NO_2 at 464.9 nm in the time-gated excitation spectrum. Chem. Phys. Lett. **99**, 475 (1983)

Sugimoto, N., Takezawa, S.: Rotational analyses and vibrational assignments of the 463- and 474-nm bands of NO_2. J. Mol. Spectrosc. **106**, 307 (1984)

Sugimoto, N., Takezawa, S., Takeuchi, N.: Rotational analysis and radiative lifetime measurement on the $^2B_1(K' = 0)$ excited state of NO_2 with $v_2' = 6, 7, 8$, and 9. J. Mol. Spectrosc. **102**, 372 (1983)

Suzuki, T.: Optogalvanic spectroscopy with rf discharge. Opt. Commun. **38**, 364 (1981)

Takezawa, S., Sugimoto, N., Takeuchi, N.: Rotational analysis of the $^2B_1$ $(K' = 0) \leftarrow {}^2A_1(K'' = 0)$ band with $v_2' = 6, 7$ and 10 in the excitation spectrum of NO_2. Chem. Phys. Lett. **97**, 77 (1983)

Teets, R.E., Carlson, N.W., Schawlow, A.L.: Polarization labeling spectroscopy of NO_2. J. Mol. Spectrosc. **78**, 415 (1979)

Vedder, H.J., Schwarz, M., Foth, H.-J., Demtröder, W.: Analysis of the perturbed NO_2 $^2B_2 \leftarrow {}^2A_1$ system in the 591.4- to 592.9-nm region based on sub-Doppler laser spectroscopy. J. Mol. Spectrosc. **97**, 92 (1983)

4.5 Double Resonance and Level-Crossing Spectroscopy

Allegrini, M., McKellar, A.R.W., Pinson, P., Brown, J.M.: Spin-rotation and hyperfine parameters for the (0 0 1) excited vibrational state of NO_2. J. Chem. Phys. **73**, 6086 (1980)

Bonilla, I.R., Demtröder, W.: Level crossing spectroscopy of NO_2 using Doppler reduced laser excitation in molecular beam. Chem. Phys. Lett. **53**, 223 (1978)

Brown, J.M., Steimle, T.C., Coles, M.E., Curl, R.F., Jr.: A determination of the spin-rotation parameters for NO_2 in the $\tilde{X}^2A_1$ state by microwave-optical double resonace. J. Chem. Phys. **74**, 3668 (1981)

Bylicki, F., König, F. Weber, H.G.: Magnetic field decoupling experiments in NO_2. Chem. Phys. Lett. **88**, 142 (1982)

Bylicki, F., Kolwas, M., Weber, H.G.: Optical saturation resonances in radio-frequency-dressed molecules. Phys. Rev. A **25**, 1550 (1982)

Bylicki, F., Weber, H.G.: Optical-radio frequency two-photon transitions in the double resonance spectrum of NO_2. Phys. Lett. A **82A**, 456 (1981)

Bylicki, F., Weber, H.G.: A very narrow level structure in NO_2: decoupling and level-cross spectroscopy in NO_2. Chem. Phys. Lett. **79**, 355 (1981)

Bylicki, F., Weber, H.G.: An irregular g factor in the NO_2 $^2B_2$ state: optical-radiofrequency double-resonance and zero-field level-crossing spectroscopy on NO_2. Chem. Phys. Lett. **79**, 517 (1981)

Bylicki, F., Weber, H.G.: Zero-field Stark level crossing and Stark-Zeeman recrossing experiments in the 593 nm band of NO_2. Chem. Phys. **70**, 299 (1982)

Bylicki, F., Weber, H.G.: Magnetic resonance experiments and a new molecular property in NO_2. J. Chem. Phys. **78**, 2899 (1983)

Herman, I.P., Javan, A., Field, R.W.: Observation of infrared-optical double resonance in NO_2. J. Chem. Phys. **68**, 2398 (1978)

Khoobehi, B., Roberts, J.A.: Infrared-microwave double-resonance probing of the population-depopulation of rotational states in the NO_2 molecule. J. Mol. Spectrosc. **97**, 227 (1983)

Solarz, R., Levy, D.H.: Microwave-optical double resonance in NO_2. J. Mol. Spectrosc. **73**, 374 (1978)

Weber, H.G.: Excited state dynamics in NO_2: an interpretation of Hanle and double resonance experiments. J. Mol. Struct. **61**, 141 (1980)

Weber, H.G., Brucat, P.J., Demtröder, W., Zare, R.N.: Measurement of NO_2 $^2B_2$ state g values by optical radio frequency double resonance. J. Mol. Spectrosc. **75**, 58 (1979)

Weber, H.G., Brucat, P.J., Zare, R.N.: Evidence for the preparation of a nonstationary state from the optical double resonance spectrum of NO_2. Chem. Phys. Lett. **63**, 217 (1979)

Weber, H.G., Bylicki, F., König, F., Burkhardt, H., Wieland, J.: A new molecular fluorescence depolarization effect: the coherence decay time of NO_2. Z. Phys. A **309**, 363 (1983)

4.6 Fluorescence Lifetimes and Cross-Section Measurements

Brucat, P.J., Zare, R.N.: NO_2 $\tilde{A}^2B_2$ state properties from Zeeman quantum beats. J. Chem. Phys. **78**, 100 (1983)

Chen, C.H., Kramer, S.D., Clark, D.W., Payne, M.G.: Studies of lifetimes of rotationally cooled NO_2 using time-resolved fluorescence excitation spectra. Chem. Phys. Lett. **65**, 419 (1979)

Cheshnovsky, O., Amirav, A.: Lifetimes of electronically excited NO_2 in a collision-free differentially pumped supersonic molecular beam. Chem. Phys. Lett. **109**, 368 (1984)

Donnelly, V.M., Kaufman, F.: Reply to comment on "Mechanism of NO_2 fluorescence quenching". J. Chem. Phys. **68**, 5671 (1978)

Donnelly, V.M., Kaufman, F.: Fluorescence lifetime studies of NO_2. II. Dependence of the perturbed $^2B_2$ state lifetimes on excitation energy. J. Chem. Phys. **69**, 1456 (1978)

Donnelly, V.M., Keil, D.G., Kaufman, F.: Fluorescence lifetime studies of NO_2. III. Mechanism of fluorescence quenching. J. Chem. Phys. **71**, 659 (1979)

Dulcey, C.S., McGee, T.J., McIlrath, T.J.: Laser-induced fluorescence in NO_2: absolute fluorescence cross sections and quantum yields. Chem. Phys. Lett. **76**, 80 (1980)

Imasaka, T., Ogawa, T., Ishibashi, N.: Inter- and intramolecular radiationless transitions of NO_2 at around 454.6 nm. Chem. Phys. **45**, 273 (1980)

Keil, D.G., Donnelly, V.M., Kaufman, F.: Fluorescence lifetime studies of NO_2. IV. Temperature dependence of fluorescence spectra and of collisional quenching of fluorescence. J. Chem. Phys. **73**, 1514 (1980)

Lee, L.C., Chiang, C.C.: Fluorescence yields from photodissociation of NO_2 at 1060–1620 Å. J. Chem. Phys. **76**, 4462 (1982)

Levy, D.H.: Comment on "Mechanism of NO_2 fluorescence quenching". J. Chem. Phys. **68**, 5669 (1978)

Matsuzaki, A., Nagakura, S.: On the mechanism of magnetic quenching of fluorescence in the gaseous state. J. Lumin. **18–19**, 115 (1979)

Nokes, C.J., Phillips, L.F.: Quenching rates for NO_2 ($\tilde{A}^2B_2$) from fluorescence decay measurements. J. Photochem. **9**, 399 (1978)
Sugimoto, N., Takezawa, S., Takeuchi, N.: Time-resolved, dispersed laser-induced fluorescence of NO_2: observation of collision-induced energy transfer effect. Jpn. J. Appl. Phys. **21**, 1536 (1982)
Takezawa, S., Sugimoto, N., Takeuchi, N.: Decay times of dispersed NO_2 fluorescence excited in the blue region. Jpn. J. Appl. Phys. **21**, 809 (1982)

4.7 Other Experimental Studies

Antonov, V.S.: Effect of the nuclear configuration of molecules in electron-excited states on the character of two-stage photoionization processes. Opt. Spektrosk. **52**, 10 (1982) [English transl.: Opt. Spectrosc. **52**, 5 (1982)]
Burak, I., Tsao, J., Prior, Y., Yablonovitch, E.: Multiphoton vibrational pumping of optically prepared NO_2 molecules. Chem. Phys. Lett. **68**, 31 (1979)
Feldmann, D., Zacharias, H., Welge, K.H.: IR multiple-photon fluorescence excitation and dissociation of electronically excited NO_2. Chem. Phys. Lett. **69**, 466 (1980)
Hodges, R.V., Lee, L.C., Moseley, J.T.: Multiphoton ionization of xenon and several small molecules at 193 and 248 nm. Int. J. Mass Spectrom. Ion Phys. **39**, 133 (1981)
Imasaka, T., Shimanoe, K., Ishibashi, N.: Thermal lens spectroscopic studies of radiationless processes in NO_2. J. Chem. Phys. **79**, 3201 (1983)
Märk, T.D., Märk, E., Stephan, K.: Comment and addendum on "Single and double ionization of nitrogen dioxide by electron impact from threshold up to 180 eV." J. Chem. Phys. **74**, 3633 (1981)
Matsumi, Y., Murasawa, Y., Obi, K., Tanaka, I.: The Schumann-Runge diatomic oxygen emission, following visible multiphoton excitation of NO_2. Laser Chem. **1**, 113 (1983)
McKendrick, C.B., Fotakis, C., Donovan, R.J.: Laser photodissociation of NO_2 at 248 nm and production of NO ($A^2\Sigma^+ \rightarrow X^2\Pi$) fluorescence. J. Photochem. **20**, 175 (1982)
Morrison, R.J.S., Grant, E.R.: Dynamics of the two-photon photodissociation of NO_2; a molecular beam multiphoton ionization study of NO photofragment internal energy distributions. J. Chem. Phys. **77**, 5994 (1982)
Morrison, R.J.S., Rockney, B.H., Grant, E.R.: Multiphoton ionization of NO_2: spectroscopy and dynamics. J. Chem. Phys.**75**, 2643 (1981)
Radhakrishnan, G., Ng, D., Estler, R.C.: Multiphoton ionization detection of photodissociation fragments: NO from NO_2. Chem. Phys. Lett. **84**, 260 (1981)
Slanger, T.G., Bischel, W.K., Dyer, M.J.: Nascent NO vibrational distribution from 2485 Å NO_2 photodissociation. J. Chem. Phys. **79**, 2231 (1983)

Stephan, K., Helm, H., Kim, Y.B., Seykora, G., Ramler, J., Grössl, M., Märk, E., Märk, T.D.: Single and double ionization of nitrogen dioxide by electron impact from threshold up to 180 eV. J. Chem. Phys. **73**, 303 (1980)

Tanaka, I.: Sequential multiphoton excitation of simple molecules. J. Photochem. **17**, 487 (1981)

Zacharias, H., Geilhaupt, M., Meier, K., Welge, K.H.: Laser photofragment spectroscopy of the NO_2 dissociation at 337 nm. A nonstatistical decay process. J. Chem. Phys. **74**, 218 (1981)

Atlas of Absorption and Stark Modulation Spectra

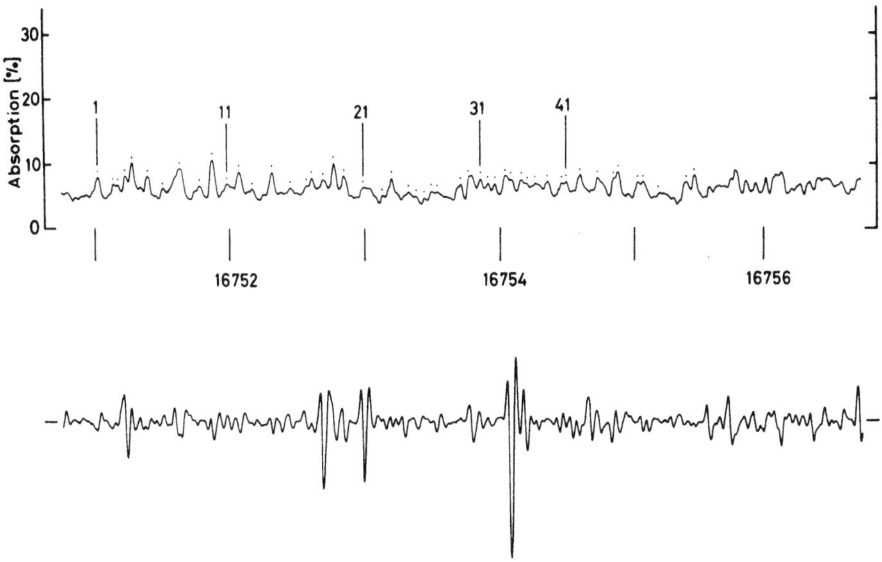

1	16751.009	21	16752.968	41	16754.470
2	16751.122	22	16753.121	42	16754.575
3	16751.156	23	16753.180	43	16754.717
4	16751.215	24	16753.316	44	16754.839
5	16751.265	25	16753.358	45	16754.869
6	16751.381	26	16753.436	46	16755.014
7	16751.497	27	16753.482	47	16755.062
8	16751.623	28	16753.521	48	16755.193
9	16751.769	29	16753.689	49	16755.391
10	16751.864	30	16753.761	50	16755.456
11	16751.976	31	16753.839		
12	16752.059	32	16753.892		
13	16752.159	33	16753.948		
14	16752.304	34	16754.027		
15	16752.446	35	16754.068		
16	16752.557	36	16754.142		
17	16752.595	37	16754.180		
18	16752.682	38	16754.248		
19	16752.755	39	16754.334		
20	16752.828	40	16754.432		

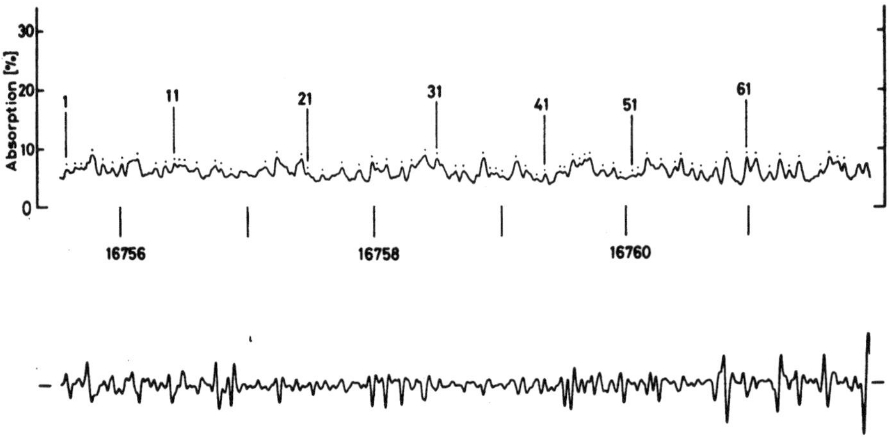

1	16755.572	21	16757.456	41	16759.337	61	16760.976
2	16755.632	22	16757.573	42	16759.454	62	16761.048
3	16755.673	23	16757.728	43	16759.497	63	16761.169
4	16755.770	24	16757.865	44	16759.557	64	16761.255
5	16755.849	25	16757.965	45	16759.609	65	16761.334
6	16755.922	26	16758.007	46	16759.650	66	16761.418
7	16755.998	27	16758.075	47	16759.691	67	16761.594
8	16756.128	28	16758.197	48	16759.800	68	16761.678
9	16756.264	29	16758.272	49	16759.889	69	16761.744
10	16756.340	30	16758.387	50	16759.960	70	16761.802
11	16756.411	31	16758.483	51	16760.043		
12	16756.450	32	16758.620	52	16760.091		
13	16756.488	33	16758.693	53	16760.168		
14	16756.577	34	16758.849	54	16760.282		
15	16756.723	35	16758.918	55	16760.394		
16	16756.769	36	16758.957	56	16760.443		
17	16756.852	37	16759.003	57	16760.528		
18	16757.112	38	16759.108	58	16760.607		
19	16757.212	39	16759.186	59	16760.723		
20	16757.398	40	16759.268	60	16760.809		

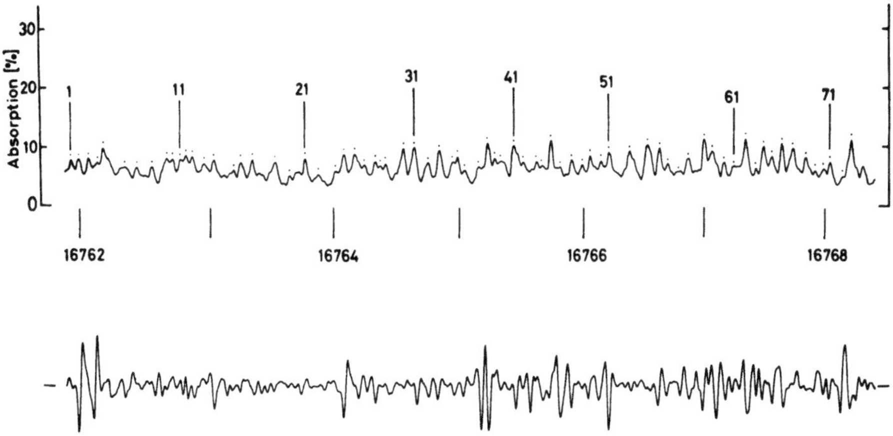

1	16761.924	21	16763.761	41	16765.433	61	16767.240
2	16761.989	22	16763.864	42	16765.510	62	16767.334
3	16762.063	23	16764.009	43	16765.616	63	16767.416
4	16762.125	24	16764.071	44	16765.667	64	16767.483
5	16762.175	25	16764.162	45	16765.729	65	16767.553
6	16762.340	26	16764.236	46	16765.796	66	16767.640
7	16762.439	27	16764.328	47	16765.896	67	16767.739
8	16762.557	28	16764.370	48	16765.985	68	16767.842
9	16762.675	29	16764.408	49	16766.045	69	16767.917
10	16762.717	30	16764.548	50	16766.138	70	16767.988
11	16762.778	31	16764.637	51	16766.199	71	16768.043
12	16762.821	32	16764.742	52	16766.369	72	16768.142
13	16762.875	33	16764.838	53	16766.512	73	16768.221
14	16762.964	34	16764.939	54	16766.613		
15	16763.037	35	16764.978	55	16766.684		
16	16763.181	36	16765.041	56	16766.778		
17	16763.248	37	16765.153	57	16766.856		
18	16763.339	38	16765.224	58	16766.995		
19	16763.524	39	16765.280	59	16767.060		
20	16763.636	40	16765.327	60	16767.159		

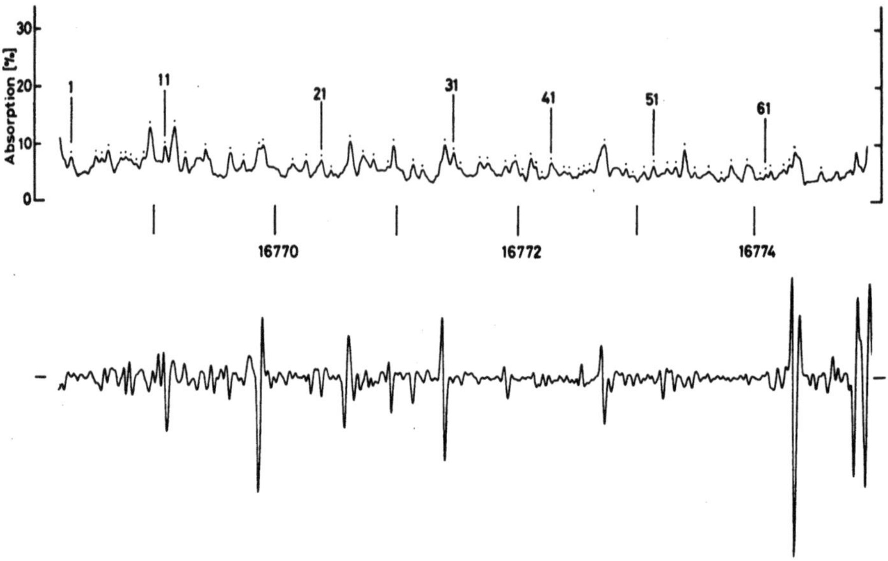

1	16768.326	21	16770.363	41	16772.272	61	16774.100
2	16768.515	22	16770.444	42	16772.379	62	16774.143
3	16768.562	23	16770.608	43	16772.421	63	16774.253
4	16768.608	24	16770.711	44	16772.503	64	16774.307
5	16768.717	25	16770.796	45	16772.547	65	16774.353
6	16768.753	26	16770.919	46	16772.598	66	16774.577
7	16768.799	27	16770.966	47	16772.717		
8	16768.839	28	16771.125	48	16772.900		
9	16768.910	29	16771.204	49	16772.958		
10	16768.961	30	16771.389	50	16773.052		
11	16769.084	31	16771.460	51	16773.135		
12	16769.160	32	16771.520	52	16773.251		
13	16769.248	33	16771.679	53	16773.324		
14	16769.411	34	16771.730	54	16773.398		
15	16769.625	35	16771.893	55	16773.484		
16	16769.729	36	16771.974	56	16773.603		
17	16769.858	37	16772.029	57	16773.712		
18	16769.889	38	16772.102	58	16773.797		
19	16770.122	39	16772.143	59	16773.934		
20	16770.236	40	16772.195	60	16774.049		

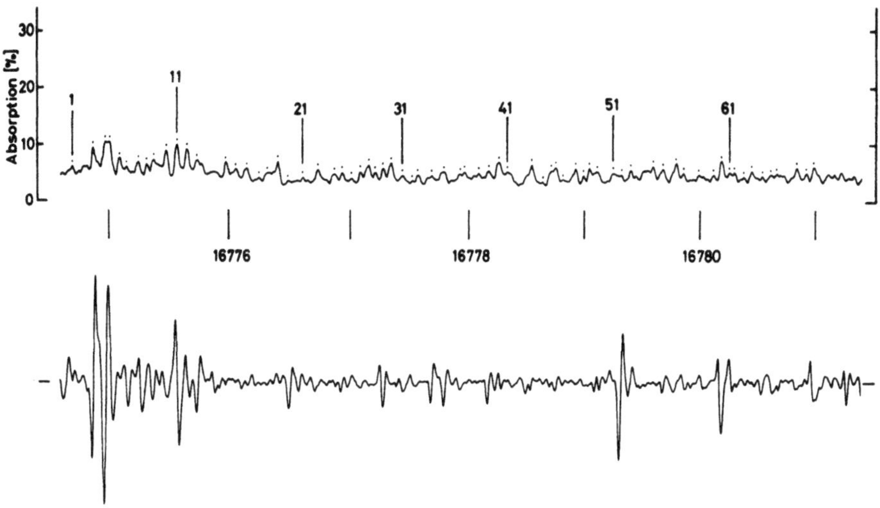

1	16774.706	21	16776.597	41	16778.317	61	16780.225
2	16774.874	22	16776.722	42	16778.524	62	16780.270
3	16774.976	23	16776.858	43	16778.635	63	16780.345
4	16775.011	24	16776.924	44	16778.697	64	16780.414
5	16775.093	25	16777.001	45	16778.741	65	16780.511
6	16775.150	26	16777.075	46	16778.802	66	16780.588
7	16775.247	27	16777.145	47	16778.910	67	16780.639
8	16775.316	28	16777.198	48	16778.972	68	16780.816
9	16775.377	29	16777.263	49	16779.027	69	16780.905
10	16775.476	30	16777.332	50	16779.089	70	16780.968
11	16775.565	31	16777.426	51	16779.236		
12	16775.647	32	16777.509	52	16779.292		
13	16775.728	33	16777.557	53	16779.374		
14	16775.963	34	16777.674	54	16779.558		
15	16776.044	35	16777.777	55	16779.641		
16	16776.134	36	16777.914	56	16779.763		
17	16776.236	37	16777.950	57	16779.825		
18	16776.307	38	16778.072	58	16779.967		
19	16776.394	39	16778.158	59	16780.088		
20	16776.477	40	16778.245	60	16780.158		

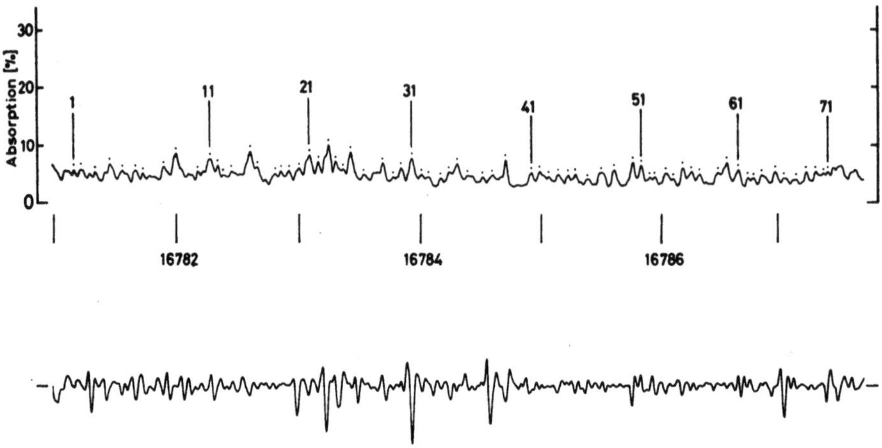

1	16781.146	21	16783.074	41	16784.898	61	16786.650
2	16781.209	22	16783.147	42	16784.968	62	16786.730
3	16781.321	23	16783.232	43	16785.042	63	16786.779
4	16781.443	24	16783.293	44	16785.124	64	16786.852
5	16781.541	25	16783.357	45	16785.209	65	16786.968
6	16781.658	26	16783.413	46	16785.269	66	16787.046
7	16781.718	27	16783.520	47	16785.369	67	16787.154
8	16781.882	28	16783.669	48	16785.483	68	16787.238
9	16781.983	29	16783.751	49	16785.595	69	16787.320
10	16782.161	30	16783.816	50	16785.754	70	16787.386
11	16782.269	31	16783.903	51	16785.825	71	16787.425
12	16782.324	32	16783.989	52	16785.899		
13	16782.384	33	16784.042	53	16785.942		
14	16782.466	34	16784.153	54	16786.034		
15	16782.594	35	16784.216	55	16786.099		
16	16782.656	36	16784.277	56	16786.179		
17	16782.788	37	16784.359	57	16786.253		
18	16782.851	38	16784.494	58	16786.322		
19	16782.910	39	16784.575	59	16786.476		
20	16782.988	40	16784.681	60	16786.551		

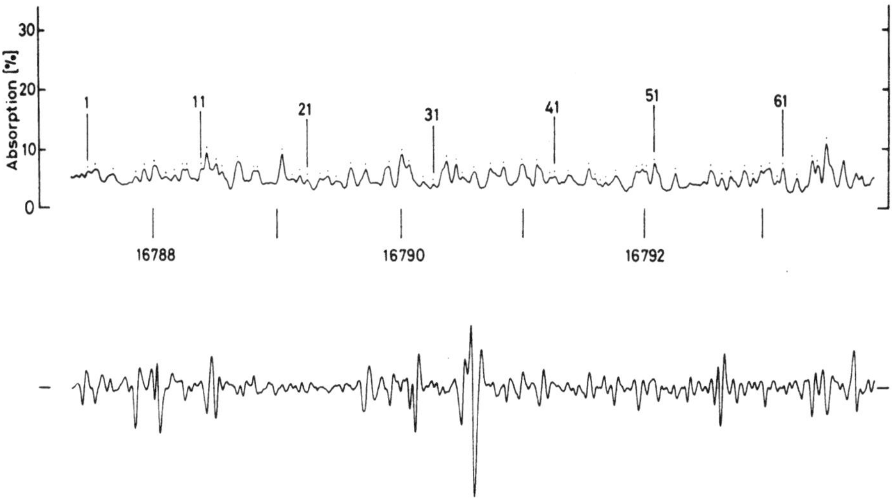

1	16787.473	21	16789.243	41	16791.253	61	16793.170
2	16787.517	22	16789.344	42	16791.371	62	16793.291
3	16787.658	23	16789.399	43	16791.535	63	16793.430
4	16787.849	24	16789.472	44	16791.589	64	16793.476
5	16787.914	25	16789.589	45	16791.636	65	16793.542
6	16788.007	26	16789.709	46	16791.677		
7	16788.105	27	16789.895	47	16791.762		
8	16788.163	28	16790.003	48	16781.933		
9	16788.227	29	16790.066	49	16791.978		
10	16788.262	30	16790.180	50	16792.013		
11	16788.376	31	16790.267	51	16792.081		
12	16788.419	32	16790.370	52	16792.252		
13	16788.492	33	16790.447	53	16792.561		
14	16788.540	34	16790.503	54	16792.657		
15	16788.674	35	16790.592	55	16792.730		
16	16788.804	36	16790.726	56	16792.846		
17	16788.840	37	16790.834	57	16792.920		
18	16789.035	38	16790.986	58	16792.985		
19	16789.125	39	16791.106	59	16793.059		
20	16789.178	40	16791.215	60	16793.123		

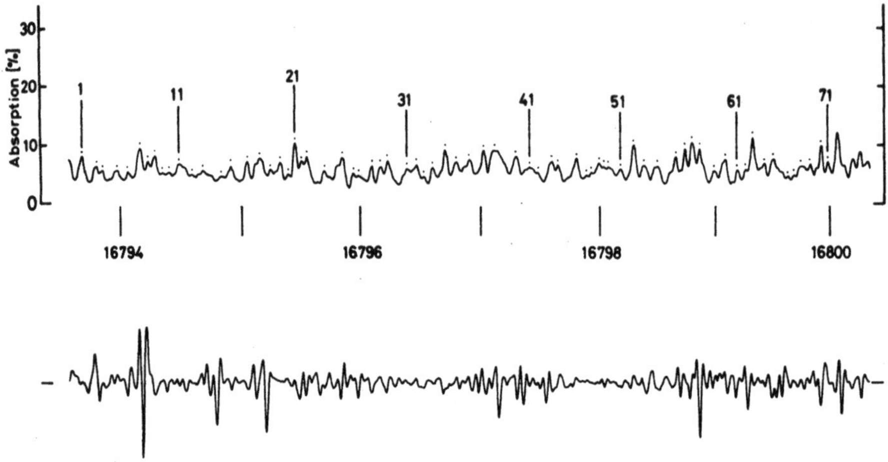

1	16793.688	21	16795.452	41	16797.406	61	16799.174
2	16793.795	22	16795.504	42	16797.478	62	16799.250
3	16793.844	23	16795.546	43	16797.594	63	16799.312
4	16793.969	24	16795.691	44	16797.649	64	16799.413
5	16794.061	25	16795.847	45	16797.803	65	16799.497
6	16794.164	26	16795.935	46	16797.888	66	16799.624
7	16794.226	27	16795.976	47	16797.928	67	16799.680
8	16794.283	28	16796.088	48	16797.991	68	16799.748
9	16794.347	29	16796.162	49	16798.035	69	16799.824
10	16794.397	30	16796.224	50	16798.073	70	16799.919
11	16794.482	31	16796.385	51	16798.170	71	16799.976
12	16794.586	32	16796.460	52	16798.287		
13	16794.668	33	16796.519	53	16798.373		
14	16794.826	34	16796.592	54	16798.489		
15	16794.909	35	16796.702	55	16798.643		
16	16795.044	36	16796.794	56	16798.724		
17	16795.153	37	16796.899	57	16798.787		
18	16795.241	38	16797.018	58	16798.848		
19	16795.324	39	16797.120	59	16798.972		
20	16795.387	40	16797.288	60	16799.071		

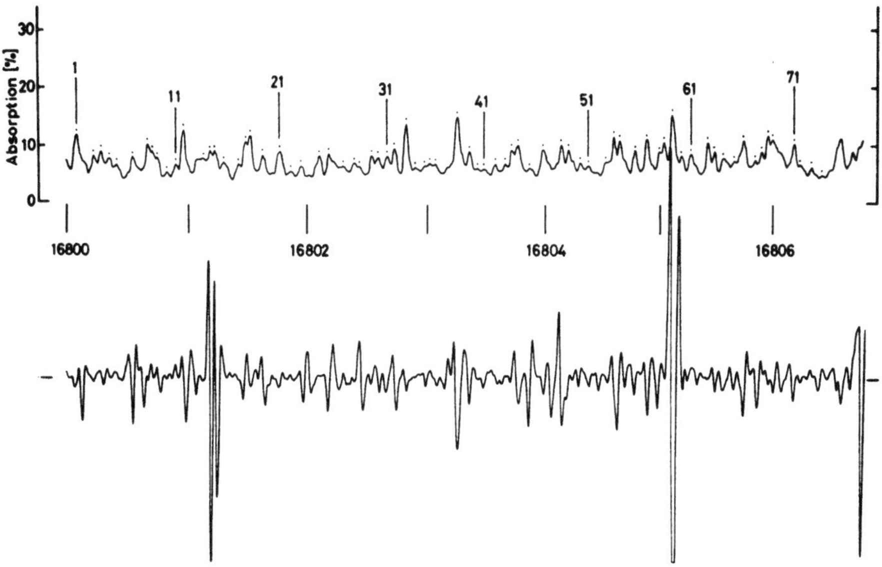

1	16800.073	21	16801.763	41	16803.467	61	16805.270
2	16800.213	22	16801.853	42	16803.561	62	16805.418
3	16800.278	23	16801.947	43	16803.644	63	16805.475
4	16800.346	24	16802.093	44	16803.700	64	16805.560
5	16800.404	25	16802.171	45	16803.752	65	16805.643
6	16800.537	26	16802.294	46	16803.847	66	16805.725
7	16800.663	27	16802.388	47	16803.967	67	16805.828
8	16800.702	28	16802.428	48	16804.128	68	16805.886
9	16800.746	29	16802.530	49	16804.189	69	16805.941
10	16800.822	30	16802.585	50	16804.297	70	16805.983
11	16800.898	31	16802.662	51	16804.368	71	16806.181
12	16800.958	32	16802.722	52	16804.522	72	16806.235
13	16801.121	33	16802.817	53	16804.588	73	16806.336
14	16801.181	34	16802.893	54	16804.642	74	16806.427
15	16801.219	35	16802.973	55	16804.774		
16	16801.290	36	16803.010	56	16804.878		
17	16801.417	37	16803.058	57	16804.984		
18	16801.478	38	16803.239	58	16805.026		
19	16801.519	39	16803.340	59	16805.100		
20	16801.621	40	16803.407	60	16805.184		

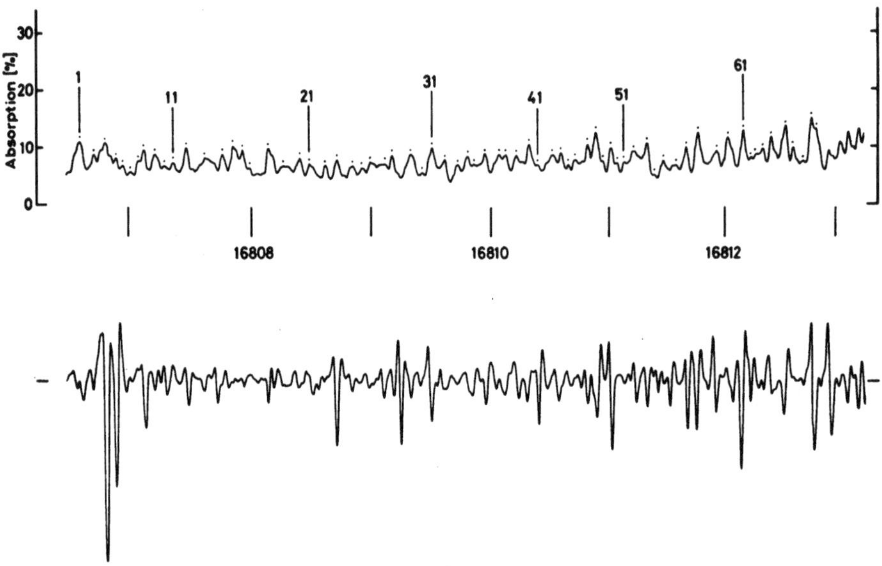

1	16806.597	21	16808.494	41	16810.390	61	16812.163
2	16806.705	22	16808.627	42	16810.519	62	16812.235
3	16806.796	23	16808.720	43	16810.587	63	16812.279
4	16806.891	24	16808.841	44	16810.654	64	16812.345
5	16806.948	25	16808.930	45	16810.712	65	16812.420
6	16807.013	26	16808.996	46	16810.821	66	16812.549
7	16807.078	27	16809.119	47	16810.888	67	16812.618
8	16807.125	28	16809.177	48	16810.949	68	16812.707
9	16807.215	29	16809.331	49	16811.017	69	16812.786
10	16807.292	30	16809.428	50	16811.068	70	16812.825
11	16807.363	31	16809.515	51	16811.132		
12	16807.468	32	16809.616	52	16811.220		
13	16807.619	33	16809.725	53	16811.333		
14	16807.767	34	16809.806	54	16811.386		
15	16807.859	35	16809.856	55	16811.464		
16	16807.938	36	16809.951	56	16811.582		
17	16807.999	37	16810.065	57	16811.671		
18	16808.149	38	16810.122	58	16811.770		
19	16808.272	39	16810.208	59	16811.934		
20	16808.408	40	16810.317	60	16812.028		

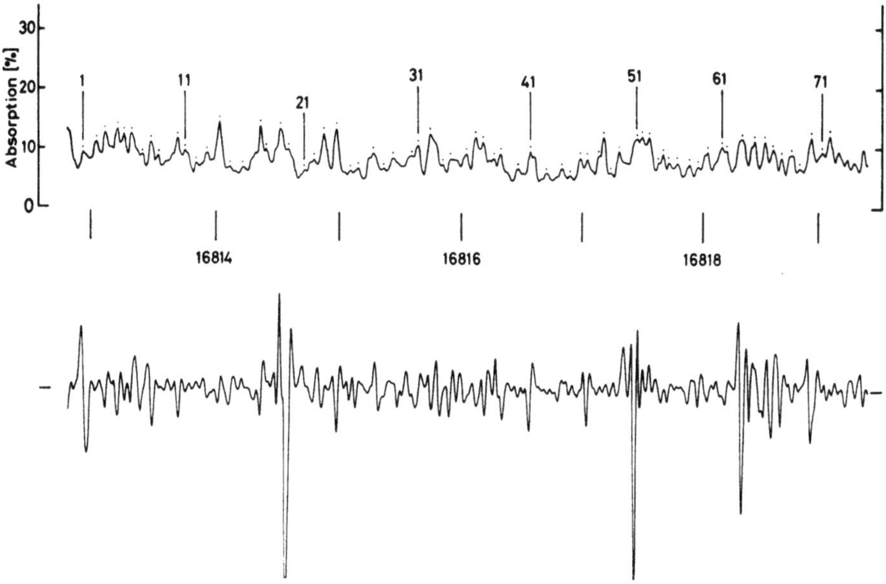

1	16812.943	21	16814.706	41	16816.553	61	16818.156
2	16813.045	22	16814.782	42	16816.691	62	16818.194
3	16813.119	23	16814.865	43	16816.816	63	16818.317
4	16813.212	24	16814.969	44	16816.889	64	16818.428
5	16813.267	25	16815.080	45	16816.977	65	16818.519
6	16813.324	26	16815.148	46	16817.034	66	16818.592
7	16813.415	27	16815.276	47	16817.133	67	16818.666
8	16813.406	28	16815.353	48	16817.170	68	16818.753
9	16813.542	29	16815.437	49	16817.230	68	16818.829
10	16813.701	30	16815.576	50	16817.298	70	16818.936
11	16813.757	31	16815.634	51	16817.442	71	16819.026
12	16813.848	32	16815.738	52	16817.488	72	16819.094
13	16813.927	33	16815.845	53	16817.545		
14	16814.028	34	16815.907	54	16817.623		
15	16814.107	35	16816.033	55	16817.662		
16	16814.215	36	16816.111	56	16817.713		
17	16814.351	37	16816.170	57	16817.781		
18	16814.396	38	16816.257	58	16817.884		
19	16814.518	39	16816.308	59	16817.960		
20	16814.576	40	16816.454	60	16818.039		

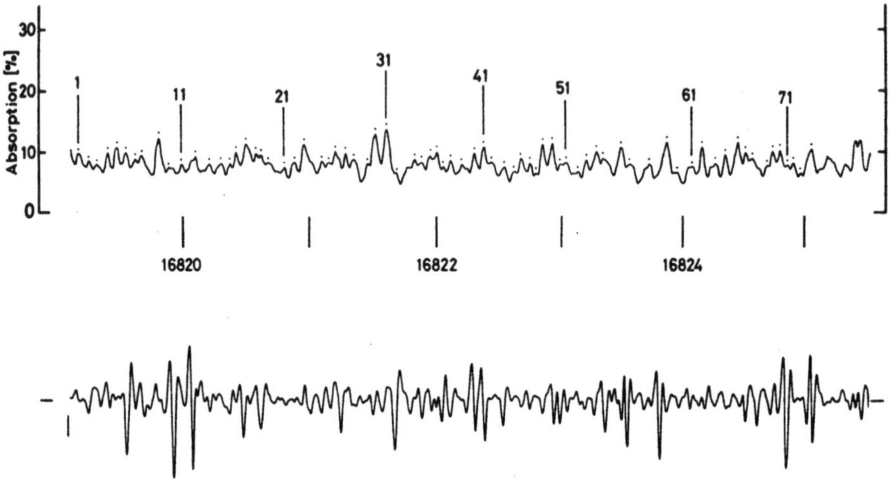

1	16819.169	21	16820.816	41	16822.382	61	16824.086
2	16819.243	22	16820.892	42	16822.441	62	16824.169
3	16819.307	23	16820.966	43	16822.543	63	16824.278
4	16819.389	24	16821.104	44	16822.619	64	16824.357
5	16819.462	25	16821.157	45	16822.677	65	16824.461
6	16819.537	26	16821.212	46	16822.748	66	16824.525
7	16819.611	27	16821.294	47	16822.799	67	16824.581
8	16819.663	28	16821.358	48	16822.853	68	16824.702
9	16819.802	29	16821.466	49	16822.932	69	16824.749
10	16819.883	30	16821.528	50	16822.999	70	16824.811
11	16819.985	31	16821.613	51	16823.044	71	16824.874
12	16820.091	32	16821.686	52	16823.141	72	16824.923
13	16820.204	33	16821.830	53	16823.211	73	16824.983
14	16820.290	34	16821.885	54	16823.295	74	16825.074
15	16820.369	35	16821.953	55	16823.348		
16	16820.421	36	16822.007	56	16823.497		
17	16820.500	37	16822.055	57	16823.567		
18	16820.582	38	16822.115	58	16823.732		
19	16820.620	39	16822.201	59	16823.877		
20	16820.687	40	16822.306	60	16823.962		

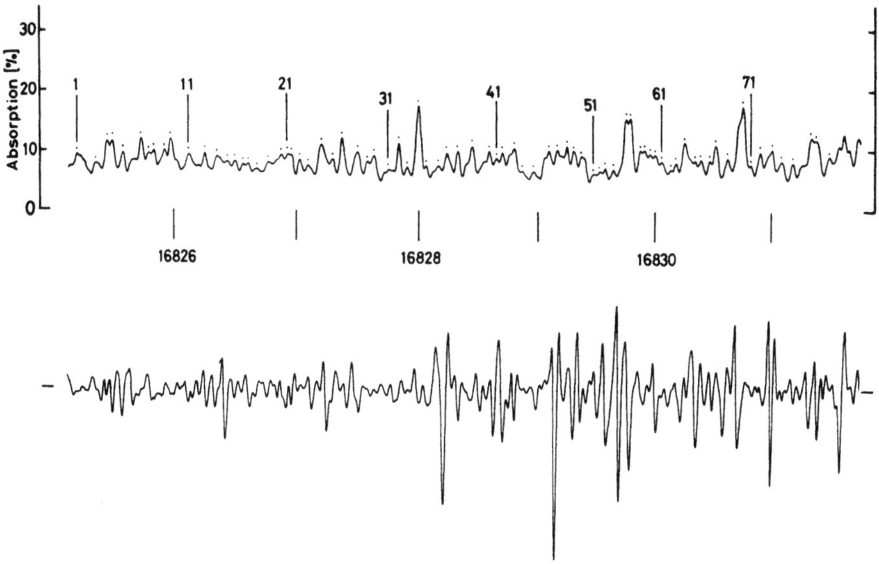

1	16825.199	21	16826.917	41	16828.645	61	16830.046
2	16825.352	22	16826.951	42	16828.696	62	16830.131
3	16825.437	23	16827.014	43	16828.795	63	16830.177
4	16825.480	24	16827.082	44	16828.870	64	16830.255
5	16825.565	25	16827.186	45	16828.963	65	16830.346
6	16825.714	26	16827.282	46	16829.091	66	16830.395
7	16825.775	27	16827.356	47	16829.161	67	16830.470
8	16825.815	28	16827.480	48	16829.243	68	16830.519
9	16825.902	29	16827.562	49	16829.307	69	16830.620
10	16825.954	30	16827.613	50	16829.369	70	16830.748
11	16826.106	31	16827.742	51	16829.468	71	16830.819
12	16826.234	32	16827.823	52	16829.522	72	16830.896
13	16826.333	33	16827.892	53	16829.565	73	16831.000
14	16826.419	34	16827.985	54	16829.630	74	16831.088
15	16826.488	35	16828.051	55	16829.735	75	16831.195
16	16826.549	36	16828.153	56	16829.772	76	16831.350
17	16826.597	37	16828.222	57	16829.861	77	16831.391
18	16826.670	38	16828.316	58	16829.899	78	16831.505
19	16826.761	39	16828.438	59	16829.945		
20	16826.867	40	16828.585	60	16829.995		

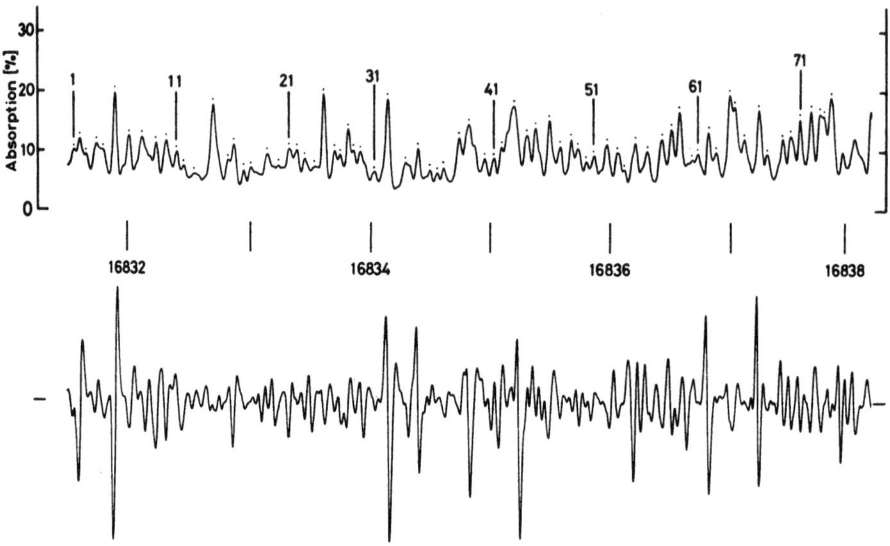

1	16831.588	21	16833.325	41	16835.015	61	16836.723
2	16831.634	22	16833.385	42	16835.077	62	16836.812
3	16831.684	23	16833.448	43	16835.179	63	16836.876
4	16831.765	24	16833.526	44	16835.290	64	16836.996
5	16831.816	25	16833.604	45	16835.361	65	16837.041
6	16831.913	26	16833.695	46	16835.482	66	16837.119
7	16832.033	27	16833.740	47	16835.576	67	16837.249
8	16832.137	28	16833.808	48	16835.667	68	16837.317
9	16832.251	29	16833.850	49	16835.725	69	16837.452
10	16832.333	30	16833.906	50	16835.796	70	16837.523
11	16832.416	31	16834.023	51	16835.860	71	16837.605
12	16832.471	32	16834.131	52	16835.964	72	16837.700
13	16832.552	33	16834.276	53	16836.056	73	16837.781
14	16832.704	34	16834.378	54	16836.116	74	16837.811
15	16832.831	35	16834.478	55	16836.203	75	16837.873
16	16832.879	36	16834.538	56	16836.305		
17	16832.959	37	16834.590	57	16836.423		
18	16833.018	38	16834.720	58	16836.505		
19	16833.148	39	16834.800	59	16836.573		
20	16833.237	40	16834.936	60	16836.677		

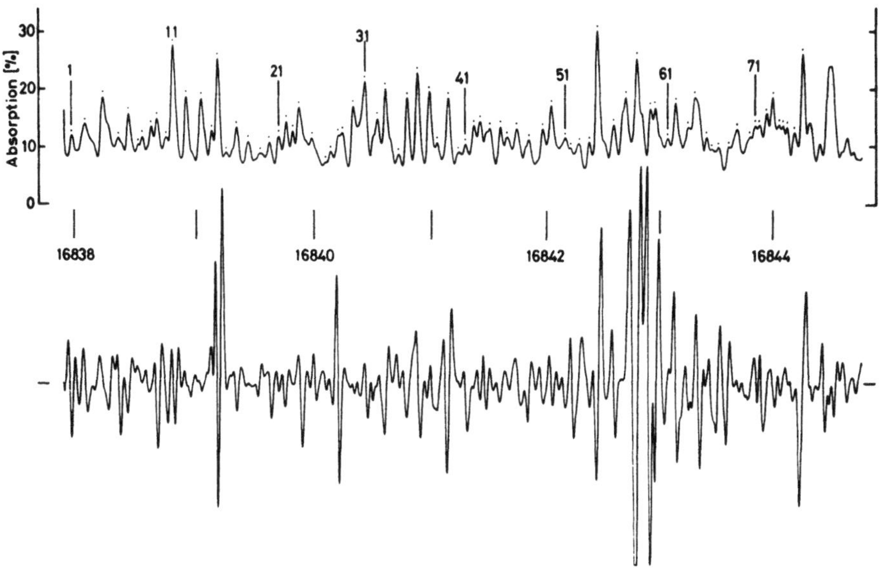

1	16837.970	21	16839.698	41	16841.284	61	16843.060
2	16838.075	22	16839.762	42	16841.360	62	16843.134
3	16838.226	23	16839.816	43	16841.415	63	16843.245
4	16838.352	24	16839.872	44	16841.498	64	16843.312
5	16838.433	25	16839.979	45	16841.591	65	16843.410
6	16838.506	26	16840.091	46	16841.650	66	16843.466
7	16838.545	27	16840.140	47	16841.732	67	16843.513
8	16838.616	28	16840.205	48	16841.838	68	16843.604
9	16838.662	29	16840.241	49	16841.960	69	16843.681
10	16838.742	30	16840.339	50	16842.035	70	16843.789
11	16838.804	31	16840.438	51	16842.151	71	16843.840
12	16838.912	32	16840.543	52	16842.202	72	16843.882
13	16839.039	33	16840.615	53	16842.270	73	16843.946
14	16839.123	34	16840.720	54	16842.366	74	16844.002
15	16839.179	35	16840.796	55	16842.442	75	16844.060
16	16839.248	36	16840.882	56	16842.583	76	16844.096
17	16839.339	37	16840.984	57	16842.693	77	16844.138
18	16839.437	38	16841.045	58	16842.791	78	16844.206
19	16839.539	39	16841.141	59	16842.911	79	16844.285
20	16839.621	40	16841.233	60	16842.953		

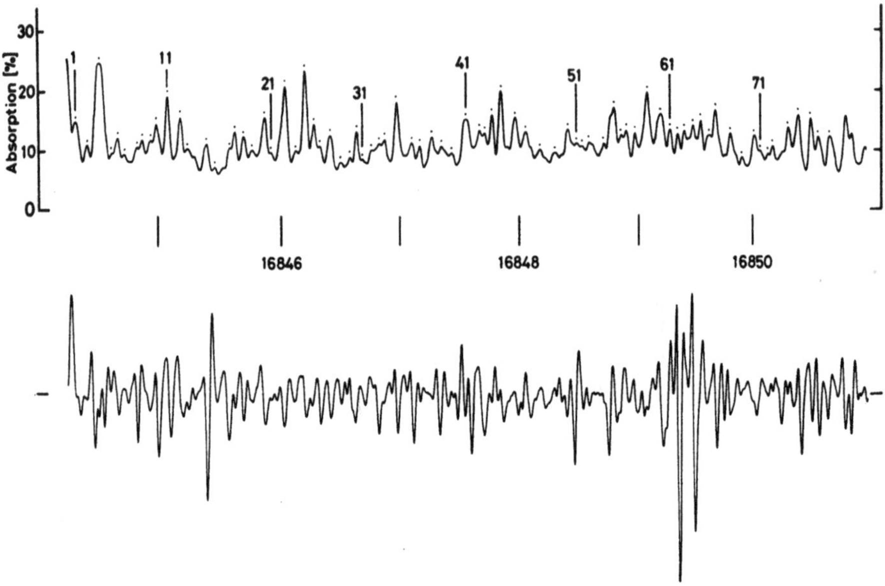

1	16844.346	21	16845.922	41	16847.547	61	16849.275
2	16844.435	22	16846.038	42	16847.664	62	16849.345
3	16844.528	23	16846.124	43	16847.704	63	16849.402
4	16844.631	24	16846.202	44	16847.767	64	16849.477
5	16844.679	25	16846.276	45	16847.840	65	16849.541
6	16844.732	26	16846.326	46	16847.958	66	16849.616
7	16844.835	27	16846.413	47	16848.046	67	16849.672
8	16844.876	28	16846.501	48	16848.168	68	16849.805
9	16844.943	29	16846.578	49	16848.306	69	16849.906
10	16844.989	30	16846.634	50	16848.402	70	16850.016
11	16845.078	31	16846.681	51	16848.474	71	16850.076
12	16845.181	32	16846.758	52	16848.533	72	16850.134
13	16845.237	33	16846.823	53	16848.583	73	16850.176
14	16845.392	34	16846.868	54	16848.709	74	16850.310
15	16845.459	35	16846.969	55	16848.794	75	16850.395
16	16845.584	36	16847.094	56	16848.859	76	16850.505
17	16845.627	37	16847.161	57	16848.902	77	16850.570
18	16845.700	38	16847.265	58	16848.976	78	16850.675
19	16845.767	39	16847.344	59	16849.081		
20	16845.873	40	16847.428	60	16849.196		

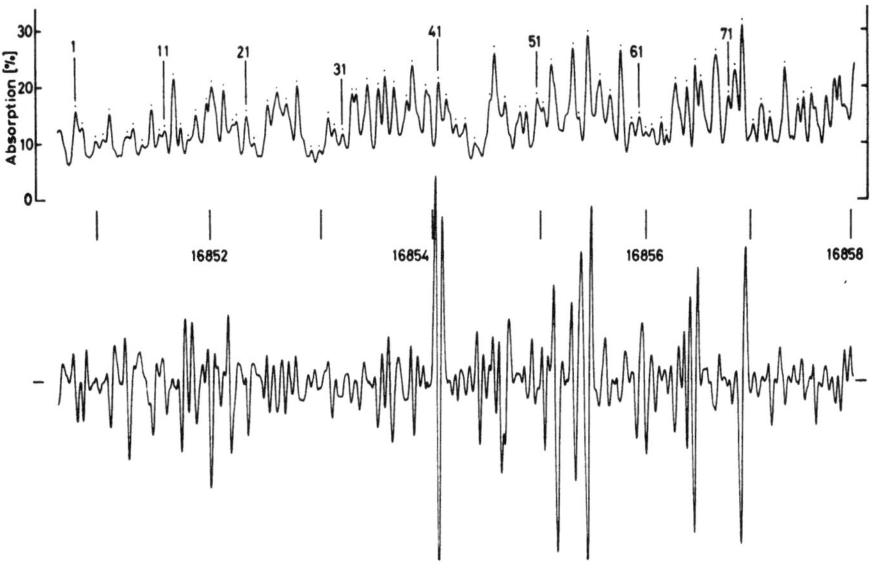

1	16850.817	21	16852.327	41	16854.057	61	16855.930
2	16850.872	22	16852.394	42	16854.131	62	16855.994
3	16850.991	23	16852.514	43	16854.218	63	16856.053
4	16851.057	24	16852.603	44	16854.307	64	16856.141
5	16851.112	25	16852.686	45	16854.395	65	16856.192
6	16851.261	26	16852.783	46	16854.531	66	16856.281
7	16851.322	27	16852.909	47	16854.573	67	16856.391
8	16851.402	28	16852.982	48	16854.670	68	16856.474
9	16851.483	29	16853.060	49	16854.810	69	16856.527
10	16851.550	30	16853.117	50	16854.865	70	16856.667
11	16851.595	31	16853.191	51	16854.971	71	16856.785
12	16851.682	32	16853.278	52	16855.026	72	16856.845
13	16851.743	33	16853.314	53	16855.100	73	16856.915
14	16851.810	34	16853.414	54	16855.308	74	16857.021
15	16851.878	35	16853.514	55	16855.365	75	16857.098
16	16851.971	36	16853.571	56	16855.452	76	16857.179
17	16852.023	37	16853.652	57	16855.567	77	16857.330
18	16852.130	38	16853.766	58	16855.662	78	16857.456
19	16852.204	39	16853.815	59	16855.764	79	16857.512
20	16852.237	40	16853.944	60	16855.864	80	16857.588

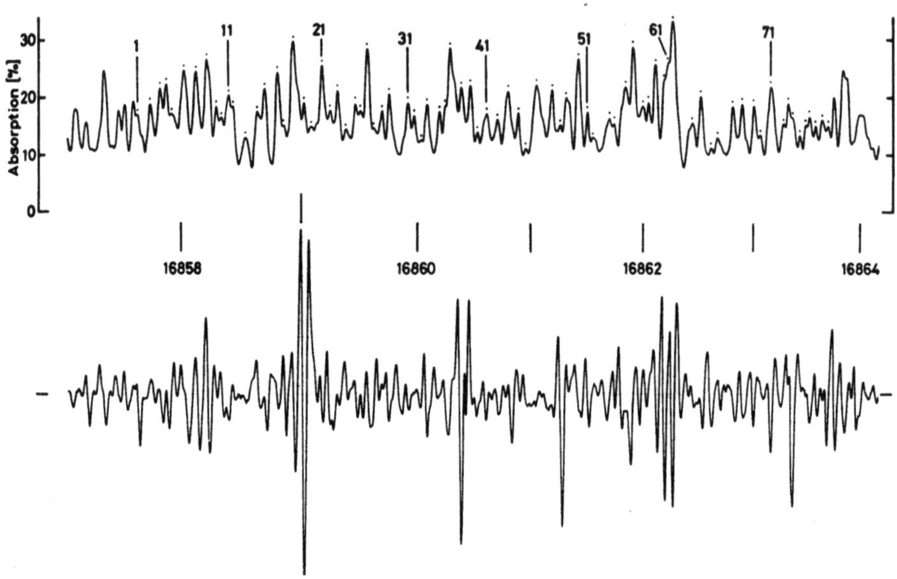

1	16857.622	21	16859.181	41	16860.612	61	16862.234
2	16857.733	22	16859.247	42	16860.704	62	16862.269
3	16857.817	23	16859.312	43	16860.802	63	16862.449
4	16857.869	24	16859.383	44	16860.886	64	16862.530
5	16857.908	25	16859.467	45	16860.955	65	16862.622
6	16858.017	26	16859.505	46	16861.050	66	16862.684
7	16858.118	27	16859.571	47	16861.191	67	16862.820
8	16858.206	28	16859.625	48	16861.267	68	16862.902
9	16858.285	29	16859.690	49	16861.323	69	16863.008
10	16858.329	30	16859.758	50	16861.433	70	16863.070
11	16858.390	31	16859.919	51	16861.515	71	16863.169
12	16858.431	32	16859.973	52	16861.567	72	16863.279
13	16858.529	33	16860.033	53	16861.708	73	16863.328
14	16858.630	34	16860.088	54	16861.751	74	16863.366
15	16858.694	35	16860.196	55	16861.848	75	16863.436
16	16858.805	36	16860.250	56	16861.916	76	16863.490
17	16858.849	37	16860.293	57	16862.001	77	16863.526
18	16858.939	38	16860.390	58	16862.055	78	16863.580
19	16859.027	39	16860.467	59	16862.115	79	16863.642
20	16859.082	40	16860.527	60	16862.196	80	16863.696

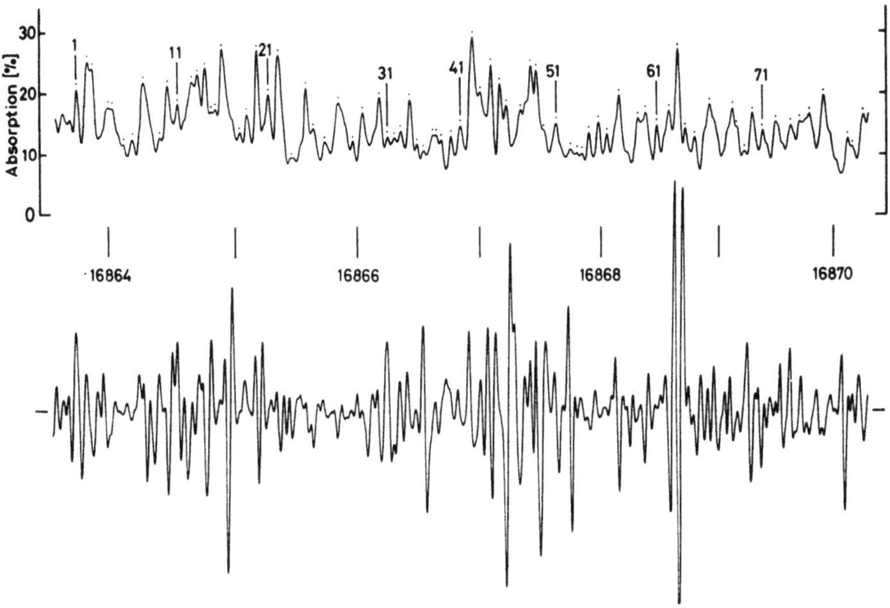

1	16863.754	21	16865.276	41	16866.841	61	16868.464
2	16863.837	22	16865.356	42	16866.944	62	16868.575
3	16863.868	23	16865.456	43	16867.005	63	16868.653
4	16863.993	24	16865.581	44	16867.100	64	16868.715
5	16864.022	25	16865.638	45	16867.171	65	16868.795
6	16864.108	26	16865.735	46	16867.222	66	16868.923
7	16864.180	27	16865.844	47	16867.339	67	16869.115
8	16864.272	28	16865.961	48	16867.428	68	16869.155
9	16864.402	29	16866.040	49	16867.473	69	16869.216
10	16864.466	30	16866.182	50	16867.534	70	16869.291
11	16864.546	31	16866.245	51	16867.629	71	16869.376
12	16864.662	32	16866.302	52	16867.744	72	16869.490
13	16864.705	33	16866.353	53	16867.801	73	16869.625
14	16864.768	34	16866.428	54	16867.841	74	16869.712
15	16864.815	35	16866.481	55	16867.898	75	16869.791
16	16864.847	36	16866.540	56	16867.979	76	16869.912
17	16864.902	37	16866.613	57	16868.047	77	16870.112
18	16865.044	38	16866.639	58	16868.152	78	16870.154
19	16865.102	39	16866.682	59	16868.311	79	16870.250
20	16865.183	40	16866.765	60	16868.374		

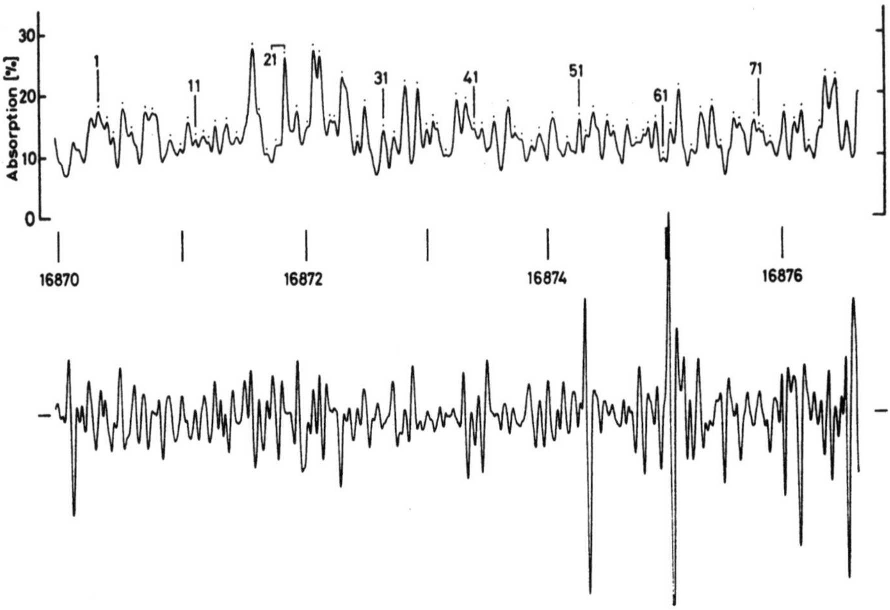

1	16870.305	21	16871.842	41	16873.390	61	16874.978
2	16870.374	22	16871.936	42	16873.453	62	16875.036
3	16870.425	23	16872.023	43	16873.555	63	16875.107
4	16870.510	24	16872.075	44	16873.673	64	16875.212
5	16870.583	25	16872.123	45	16873.728	65	16875.294
6	16870.700	26	16872.213	46	16873.781	66	16875.393
7	16870.763	27	16872.247	47	16873.860	67	16875.462
8	16870.904	28	16872.312	48	16873.929	68	16875.574
9	16870.986	29	16872.434	49	16874.041	69	16875.623
10	16871.048	30	16872.496	50	16874.168	70	16875.752
11	16871.106	31	16872.648	51	16874.266	71	16875.799
12	16871.169	32	16872.731	52	16874.324	72	16875.836
13	16871.207	33	16872.825	53	16874.384	73	16875.894
14	16871.268	34	16872.928	54	16874.499	74	16876.023
15	16871.361	35	16872.999	55	16874.551	75	16876.112
16	16871.441	36	16873.051	56	16874.669	76	16876.173
17	16871.580	37	16873.088	57	16874.746	77	16876.239
18	16871.632	38	16873.159	58	16874.810	78	16876.328
19	16871.696	39	16873.247	59	16874.845	79	16876.378
20	16871.768	40	16873.321	60	16874.911	80	16876.467

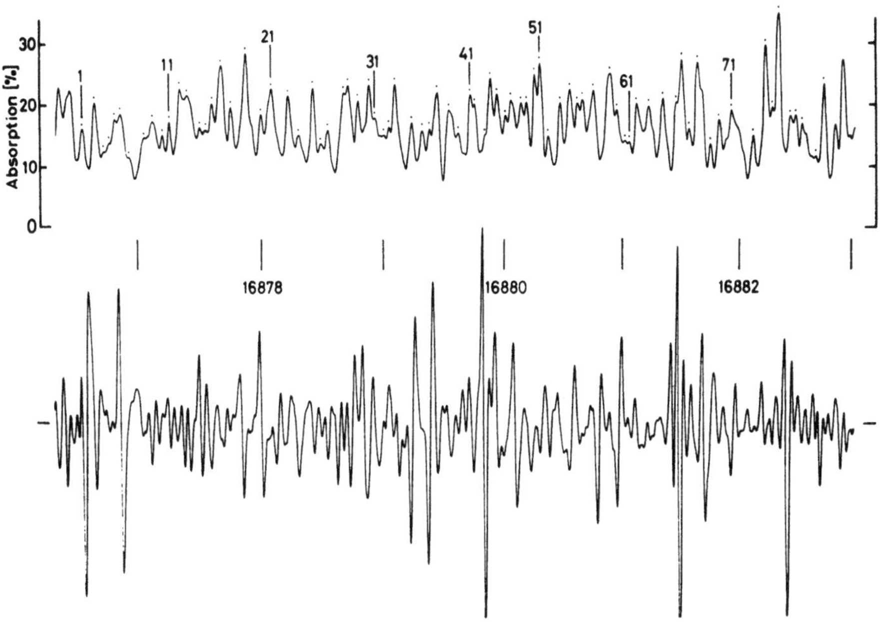

1	16876.558	21	16878.083	41	16879.729	61	16881.060
2	16876.658	22	16878.215	42	16879.764	62	16881.124
3	16876.733	23	16878.304	43	16879.848	63	16881.224
4	16876.771	24	16878.421	44	16879.895	64	16881.342
5	16876.821	25	16878.491	45	16879.951	65	16881.458
6	16876.865	26	16878.544	46	16880.021	66	16881.507
7	16876.931	27	16878.675	47	16880.070	67	16881.570
8	16877.048	28	16878.711	48	16880.150	68	16881.647
9	16877.119	29	16878.793	49	16880.196	69	16881.753
10	16877.198	30	16878.881	50	16880.268	70	16881.830
11	16877.256	31	16878.931	51	16880.313	71	16881.940
12	16877.339	32	16879.004	52	16880.380	72	16882.119
13	16877.400	33	16879.049	53	16880.483	73	16882.232
14	16877.496	34	16879.098	54	16880.569	74	16882.295
15	16877.540	35	16879.235	55	16880.631	75	16882.342
16	16877.591	36	16879.324	56	16880.674	76	16882.452
17	16877.667	37	16879.384	57	16880.767	77	16882.498
18	16877.748	38	16879.455	58	16880.898	78	16882.560
19	16877.873	39	16879.545	59	16880.959	79	16882.676
20	16878.000	40	16879.623	60	16881.023	80	16882.749

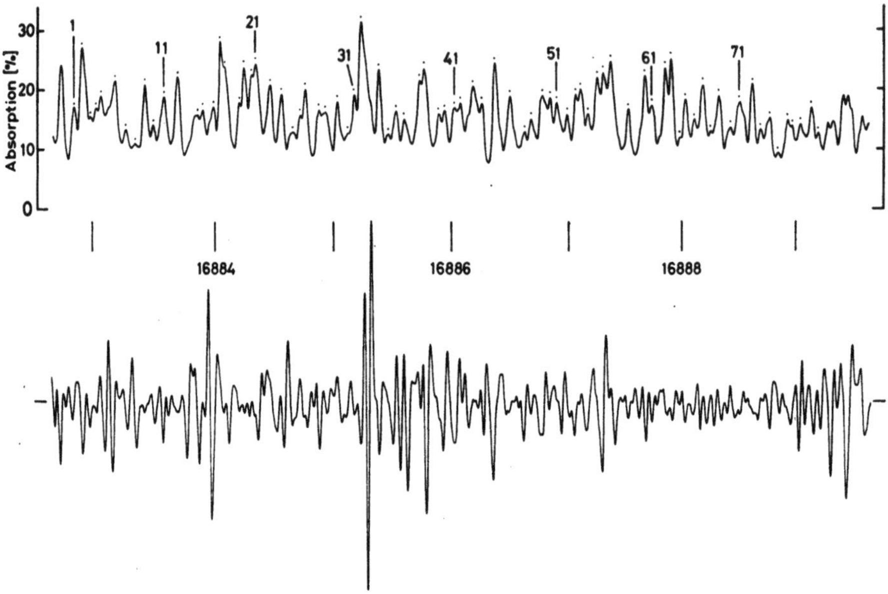

1	16882.851	21	16884.356	41	16886.021	61	16887.728
2	16882.922	22	16884.472	42	16886.072	62	16887.848
3	16882.984	23	16884.566	43	16886.180	63	16887.901
4	16883.023	24	16884.660	44	16886.254	64	16887.980
5	16883.062	25	16884.723	45	16886.367	65	16888.029
6	16883.182	26	16884.766	46	16886.496	66	16888.106
7	16883.272	27	16884.882	47	16886.622	67	16888.184
8	16883.346	28	16884.937	48	16886.674	68	16888.245
9	16883.428	29	16885.037	49	16886.773	69	16888.320
10	16883.490	30	16885.123	50	16886.841	70	16888.418
11	16883.582	31	16885.179	51	16886.896	71	16888.499
12	16883.698	32	16885.238	52	16886.985	72	16888.609
13	16883.860	33	16885.386	53	16887.061	73	16888.675
14	16883.902	34	16885.462	54	16887.102	74	16888.761
15	16883.995	35	16885.529	55	16887.164	75	16888.836
16	16884.056	36	16885.596	56	16887.248	76	16888.925
17	16884.092	37	16885.724	57	16887.302	77	16888.960
18	16884.216	38	16885.762	58	16887.366	78	16889.034
19	16884.258	39	16885.886	59	16887.518	79	16889.133
20	16884.324	40	16885.936	60	16887.668	80	16889.191

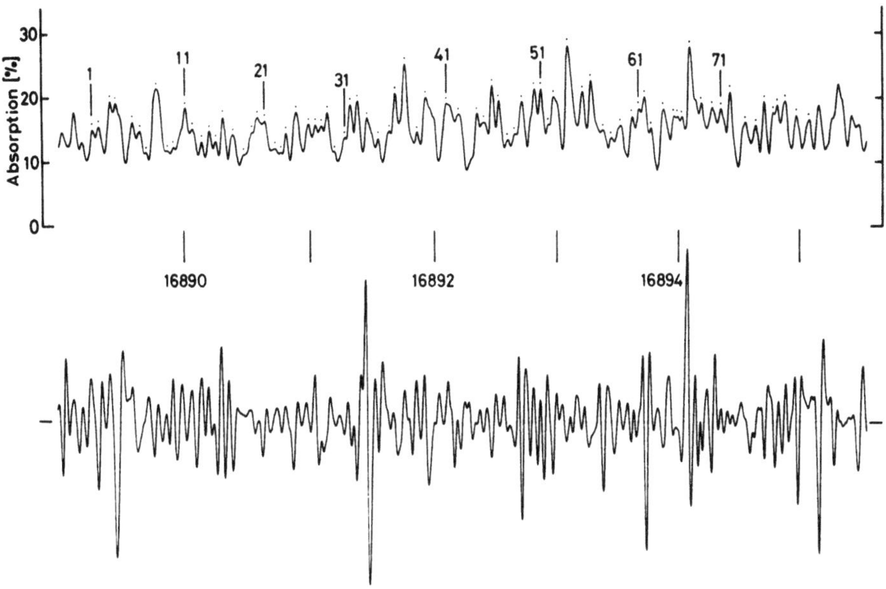

1	16889.279	21	16890.639	41	16892.092	61	16893.666
2	16889.324	22	16890.714	42	16892.191	62	16893.710
3	16889.421	23	16890.763	43	16892.342	63	16893.768
4	16889.462	24	16890.805	44	16892.405	64	16893.870
5	16889.591	25	16890.885	45	16892.468	65	16893.960
6	16889.644	26	16890.983	46	16892.522	66	16893.993
7	16889.703	27	16891.036	47	16892.589	67	16894.029
8	16889.787	28	16891.084	48	16892.648	68	16884.097
9	16889.867	29	16891.137	49	16892.710	69	16894.196
10	16889.917	30	16891.191	50	16892.814	70	16894.283
11	16890.013	31	16891.273	51	16892.868	71	16894.351
12	16890.068	32	16891.318	52	16892.922	72	16894.424
13	16890.143	33	16891.377	53	16892.972	73	16894.547
14	16890.203	34	16891.457	54	16893.083	74	16894.629
15	16890.256	35	16891.544	55	16893.204	75	16894.698
16	16890.311	36	16891.636	56	16893.274	76	16894.771
17	16890.386	37	16891.684	57	16893.371	77	16894.810
18	16890.472	38	16891.763	58	16893.436	78	16894.874
19	16890.537	39	16891.857	59	16893.514	79	16894.978
20	16890.583	40	16891.928	60	16893.609	80	16895.087

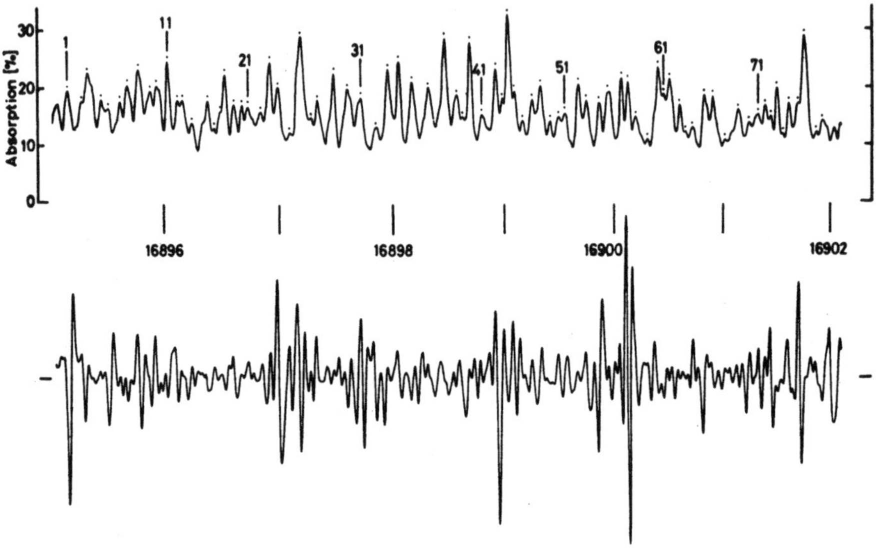

1	16895.166	21	16896.735	41	16898.797	61	16900.453
2	16895.280	22	16896.842	42	16898.919	62	16900.517
3	16895.337	23	16896.927	43	16898.979	63	16900.607
4	16895.453	24	16896.994	44	16899.033	64	16900.658
5	16895.512	25	16897.102	45	16899.090	65	16900.723
6	16895.619	26	16897.194	46	16899.164	66	16900.836
7	16895.683	27	16897.278	47	16899.250	67	16900.912
8	16895.780	28	16897.343	48	16899.327	68	16901.024
9	16895.882	29	16897.491	49	16899.417	69	16901.148
10	16895.934	30	16897.615	50	16899.497	70	16901.250
11	16896.032	31	16897.729	51	16899.549	71	16901.336
12	16896.118	32	16897.865	52	16899.675	72	16901.400
13	16896.162	33	16897.966	53	16899.747	73	16901.454
14	16896.245	34	16898.061	54	16899.859	74	16901.513
15	16896.382	35	16898.173	55	16899.942	75	16901.568
16	16896.440	36	16898.325	56	16900.064	76	16901.619
17	16896.490	37	16898.470	57	16900.126	77	16901.702
18	16896.538	38	16898.567	58	16900.199	78	16901.763
19	16896.610	39	16898.631	59	16900.311	79	16901.873
20	16896.682	40	16898.697	60	16900.409	80	16901.927

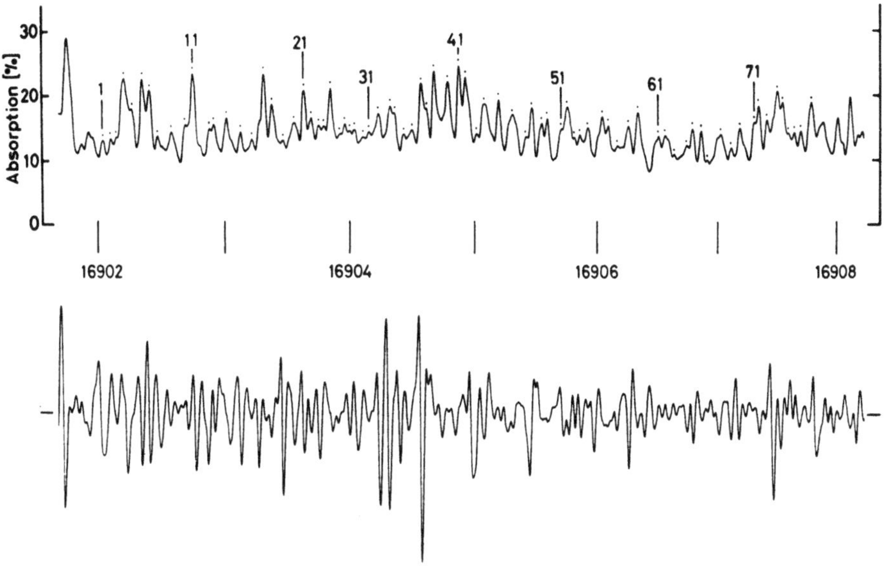

1	16902.037	21	16903.639	41	16904.875	61	16906.506
2	16902.098	22	16903.698	42	16904.930	62	16906.568
3	16902.142	23	16903.755	43	16905.014	63	16906.638
4	16902.209	24	16903.795	44	16905.085	64	16906.746
5	16902.264	25	16903.851	45	16905.198	65	16906.796
6	16902.353	26	16903.924	46	16905.306	66	16906.868
7	16902.409	27	16903.960	47	16905.412	67	16906.924
8	16902.473	28	16903.998	48	16905.467	68	16907.034
9	16902.583	29	16904.035	49	16905.543	69	16907.118
10	16902.694	30	16904.098	50	16905.594	70	16907.187
11	16902.756	31	16904.146	51	16905.711	71	16907.305
12	16902.880	32	16904.225	52	16905.756	72	16907.344
13	16902.917	33	16904.323	53	16905.816	73	16907.408
14	16903.020	34	16904.359	54	16905.860	74	16907.501
15	16903.132	35	16904.425	55	16905.924	75	16907.550
16	16903.225	36	16904.492	56	16906.049	76	16907.594
17	16903.322	37	16904.573	57	16906.097	77	16907.643
18	16903.386	38	16904.615	58	16906.174	78	16907.702
19	16903.472	39	16904.677	59	16906.262	79	16907.785
20	16903.556	40	16904.788	60	16906.342		

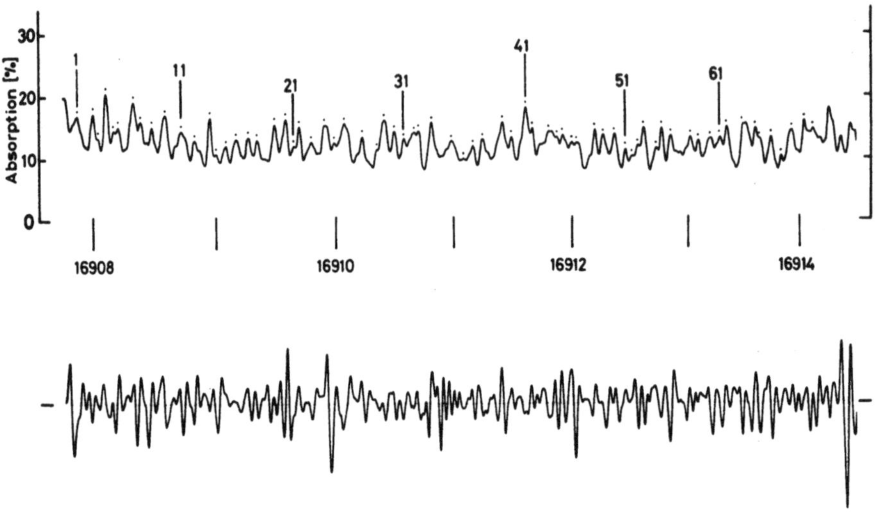

1	16907.868	21	16909.649	41	16911.609	61	16913.283
2	16907.996	22	16909.695	42	16911.662	62	16913.343
3	16908.040	23	16909.797	43	16911.802	63	16913.490
4	16908.106	24	16909.907	44	16911.869	64	16913.607
5	16908.171	25	16909.990	45	16911.927	65	16913.651
6	16908.211	26	16910.073	46	16912.000	66	16913.743
7	16908.336	27	16910.223	47	16912.049	67	16913.836
8	16908.392	28	16910.346	48	16912.204	68	16913.929
9	16908.485	29	16910.408	49	16912.276	69	16914.041
10	16908.591	30	16910.492	50	16912.373	70	16914.123
11	16908.722	31	16910.573	51	16912.471		
12	16908.829	32	16910.660	52	16912.524		
13	16908.956	33	16910.698	53	16912.576		
14	16909.011	34	16910.811	54	16912.628		
15	16909.091	35	16910.980	55	16912.729		
16	16909.177	36	16911.078	56	16912.777		
17	16909.277	37	16911.157	57	16912.852		
18	16909.348	38	16911.238	58	16913.025		
19	16909.491	39	16911.406	59	16913.094		
20	16909.583	40	16911.484	60	16913.193		

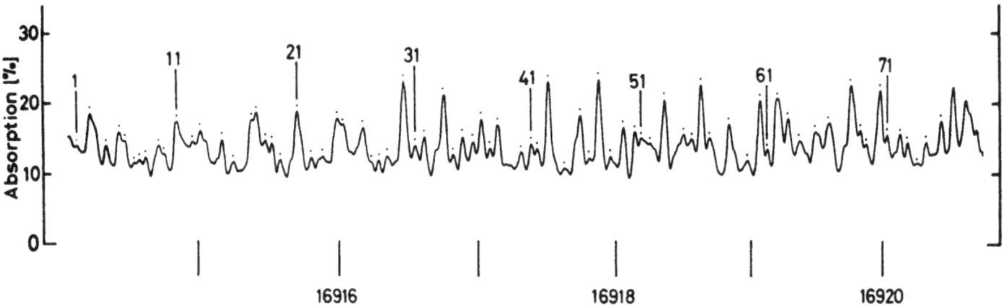

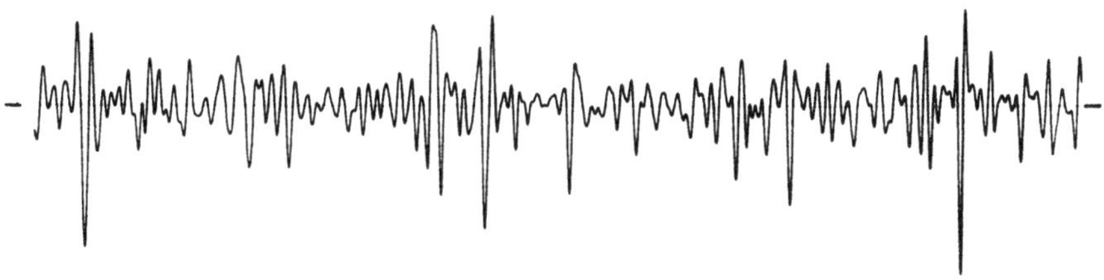

1	16914.178	21	16915.698	41	16917.383	61	16919.118
2	16914.265	22	16915.796	42	16917.427	62	16919.203
3	16914.372	23	16915.877	43	16917.506	63	16919.278
4	16914.462	24	16915.990	44	16917.614	64	16919.369
5	16914.503	25	16916.033	45	16917.734	65	16919.486
6	16914.559	26	16916.177	46	16917.802	66	16919.591
7	16914.598	27	16916.232	47	16917.873	67	16919.761
8	16914.642	28	16916.288	48	16917.958	68	16919.825
9	16914.734	29	16916.345	49	16918.050	69	16919.873
10	16914.775	30	16916.468	50	16918.139	70	16919.983
11	16914.862	31	16916.546	51	16918.192	71	16920.036
12	16914.966	32	16916.613	52	16918.248	72	16920.137
13	16915.023	33	16916.757	53	16918.359	73	16920.198
14	16915.165	34	16916.821	54	16918.501	74	16920.274
15	16915.243	35	16916.891	55	16918.557	75	16920.337
16	16915.370	36	16916.963	56	16918.630	76	16920.450
17	16915.404	37	16917.025	57	16918.697		
18	16915.468	38	16917.088	58	16918.838		
19	16915.513	39	16917.143	59	16918.976		
20	16915.571	40	16917.310	60	16919.066		

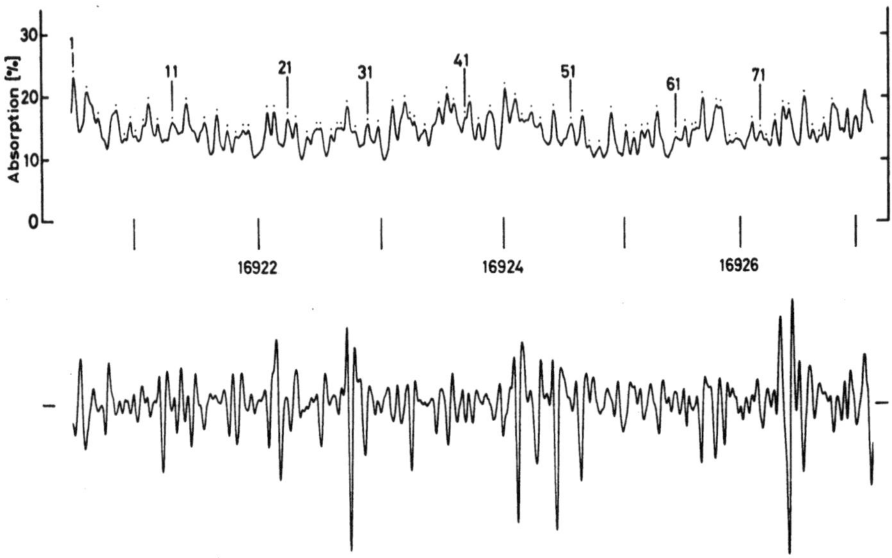

1	16920.543	21	16922.239	41	16923.687	61	16925.444
2	16920.640	22	16922.306	42	16923.726	62	16925.525
3	16920.718	23	16922.399	43	16923.796	63	16925.587
4	16920.861	24	16922.474	44	16923.887	64	16925.620
5	16920.927	25	16922.512	45	16924.014	65	16925.682
6	16920.973	26	16922.600	46	16924.096	66	16925.800
7	16921.016	27	16922.651	47	16924.146	67	16925.840
8	16921.078	28	16922.682	48	16924.228	68	16925.913
9	16921.120	29	16922.735	49	16924.296	69	16925.964
10	16921.193	30	16922.795	50	16924.410	70	16926.108
11	16921.308	31	16922.898	51	16924.554	71	16926.186
12	16921.417	32	16922.976	52	16924.649	72	16926.234
13	16921.560	33	16923.098	53	16924.704	73	16926.306
14	16921.659	34	16923.196	54	16924.788	74	16926.378
15	16921.748	35	16923.236	55	16924.893	75	16926.426
16	16921.828	36	16923.274	56	16925.016	76	16926.560
17	16921.878	37	16923.353	57	16925.082	77	16926.625
18	16921.919	38	16923.472	58	16925.148	78	16926.684
19	16922.077	39	16923.541	59	16925.201	79	16926.725
20	16922.134	40	16923.593	60	16925.286	80	16926.804

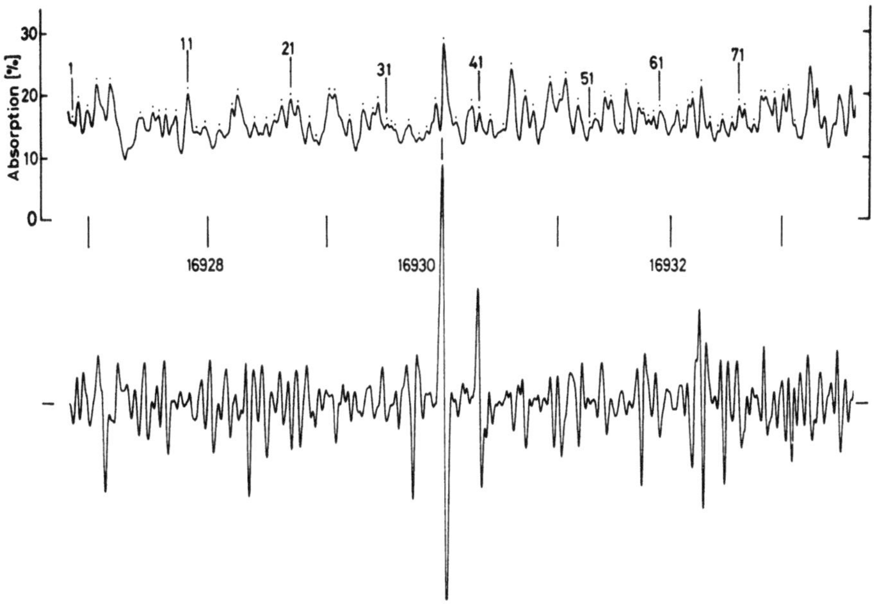

1	16926.877	21	16928.694	41	16930.326	61	16931.905
2	16926.923	22	16928.755	42	16930.418	62	16932.063
3	16926.997	23	16928.855	43	16930.523	63	16932.121
4	16927.080	24	16928.908	44	16930.603	64	16932.171
5	16927.191	25	16929.031	45	16930.721	65	16932.208
6	16927.436	26	16929.078	46	16930.791	66	16932.285
7	16927.539	27	16929.202	47	16930.943	67	16932.353
8	16927.586	28	16929.320	48	16931.030	68	16932.420
9	16927.647	29	16929.401	49	16931.078	69	16932.461
10	16927.732	30	16929.453	50	16931.186	70	16932.542
11	16927.838	31	16929.530	51	16931.291	71	16932.607
12	16927.908	32	16929.564	52	16931.340	72	16932.656
13	16927.977	33	16929.596	53	16931.425	73	16932.740
14	16928.103	34	16929.715	54	16931.477	74	16932.806
15	16928.200	35	16929.812	55	16931.561	75	16932.846
16	16928.249	36	16929.870	56	16931.614	76	16932.935
17	16928.388	37	16929.951	57	16931.728	77	16933.020
18	16928.487	38	16930.026	58	16931.765	78	16933.073
19	16928.550	39	16930.130	59	16931.810	79	16933.125
20	16928.618	40	16930.264	60	16931.852		

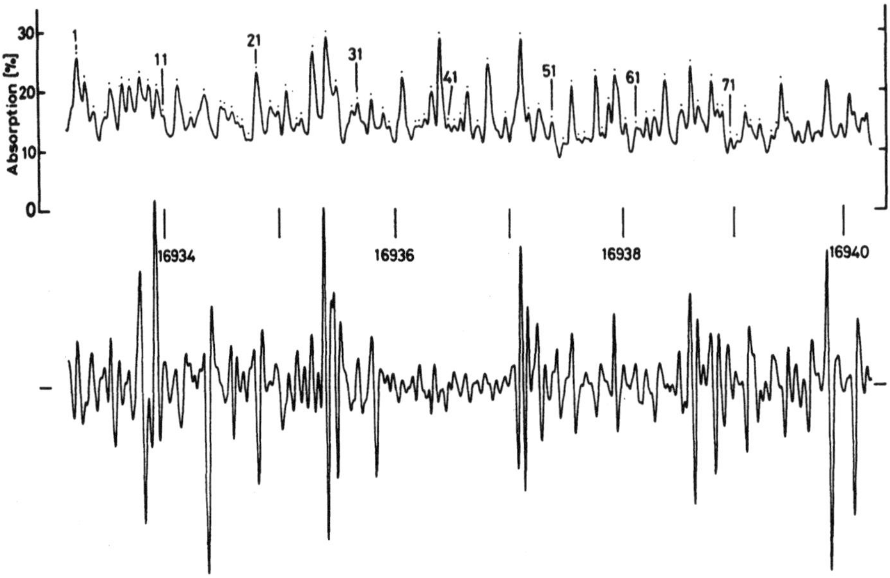

1	16933.273	21	16934.820	41	16936.457	61	16938.132
2	16933.335	22	16934.933	42	16936.508	62	16938.219
3	16933.400	23	16934.996	43	16936.564	63	16938.288
4	16933.490	24	16935.067	44	16936.623	64	16938.387
5	16933.538	25	16935.154	45	16936.712	65	16938.531
6	16933.645	26	16935.192	46	16936.808	66	16938.616
7	16933.711	27	16935.298	47	16936.953	67	16938.679
8	16933.800	28	16935.415	48	16937.092	68	16938.801
9	16933.876	29	16935.498	49	16937.155	69	16938.854
10	16933.941	30	16935.635	50	16937.239	70	16938.901
11	16934.000	31	16935.688	51	16937.352	71	16938.973
12	16934.125	32	16935.798	52	16937.459	72	16939.033
13	16934.236	33	16935.899	53	16937.533	73	16939.106
14	16934.354	34	16935.947	54	16937.595	74	16939.157
15	16934.496	35	16936.065	55	16937.653	75	16939.244
16	16934.529	36	16936.165	56	16937.753	76	16939.347
17	16934.592	37	16936.199	57	16937.810	77	16939.401
18	16934.647	38	16936.240	58	16937.866	78	16939.442
19	16934.684	39	16936.309	59	16937.928	79	16939.493
20	16934.740	40	16936.386	60	16938.027		

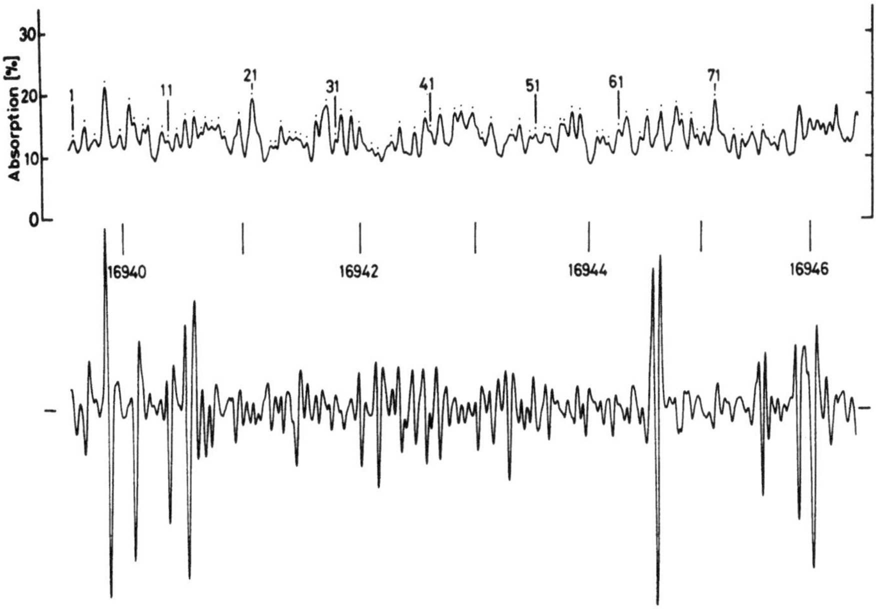

1	16939.588	21	16941.089	41	16942.610	61	16944.265
2	16939.680	22	16941.242	42	16942.692	62	16944.338
3	16939.768	23	16941.281	43	16942.822	63	16944.455
4	16939.851	24	16941.333	44	16942.879	64	16944.517
5	16939.972	25	16941.402	45	16942.977	65	16944.640
6	16940.053	26	16941.457	46	16943.062	66	16944.779
7	16940.095	27	16941.498	47	16943.138	67	16944.824
8	16940.171	28	16941.550	48	16943.318	68	16944.914
9	16940.210	29	16941.630	49	16943.385	69	16944.962
10	16940.332	30	16941.725	50	16943.476	70	16945.029
11	16940.387	31	16941.796	51	16943.521	71	16945.130
12	16940.454	32	16941.847	52	16943.611	72	16945.289
13	16940.522	33	16941.930	53	16943.650	73	16945.367
14	16940.602	34	16941.999	54	16943.748	74	16945.456
15	16940.655	35	16942.104	55	16943.781	75	16945.552
16	16940.697	36	16942.153	56	16943.851	76	16945.659
17	16940.747	37	16942.275	57	16943.923		
18	16940.796	38	16942.344	58	16944.075		
19	16940.849	39	16942.471	59	16944.125		
20	16940.974	40	16942.565	60	16944.179		

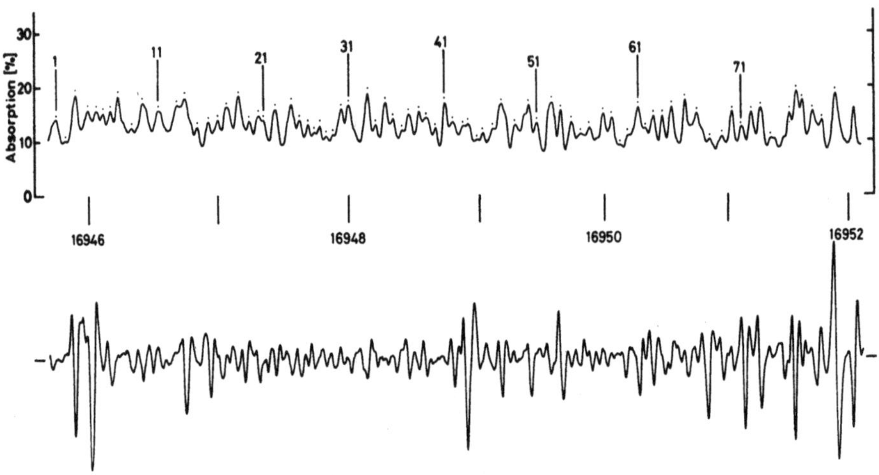

1	16945.750	21	16947.358	41	16948.743	61	16950.271
2	16945.824	22	16947.445	42	16948.806	62	16950.330
3	16945.902	23	16947.563	43	16948.890	63	16950.395
4	16945.999	24	16947.628	44	16948.925	64	16950.467
5	16946.066	25	16947.695	45	16948.998	65	16950.541
6	16946.120	26	16947.743	46	16949.042	66	16950.656
7	16946.179	27	16947.786	47	16949.117	67	16950.745
8	16946.240	28	16947.836	48	16949.185	68	16950.855
9	16946.348	29	16947.888	49	16949.294	69	16950.951
10	16946.432	30	16947.953	50	16949.400	70	16951.034
11	16946.553	31	16948.008	51	16949.465	71	16951.116
12	16946.696	32	16948.152	52	16949.584	72	16951.195
13	16946.753	33	16948.215	53	16949.654	73	16951.275
14	16946.848	34	16948.287	54	16949.739	74	16951.357
15	16946.935	35	16948.345	55	16949.814	75	16951.508
16	16947.010	36	16948.413	56	16949.882	76	16951.567
17	16947.078	37	16948.463	57	16949.991	77	16951.616
18	16947.170	38	16948.542	58	16950.060	78	16951.702
19	16947.251	39	16948.597	59	16950.130	79	16951.778
20	16947.321	40	16948.672	60	16950.194	80	16951.891

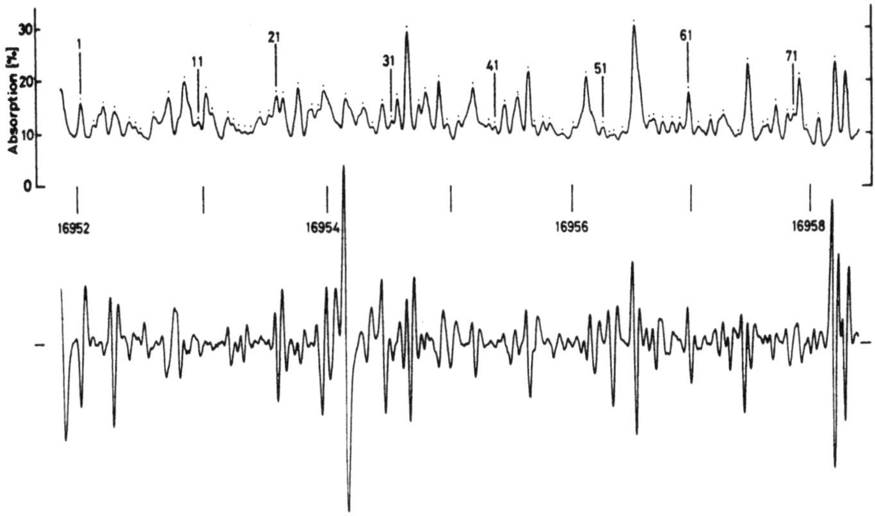

1	16952.041	21	16953.590	41	16955.359	61	16956.971
2	16952.141	22	16953.644	42	16955.435	62	16957.051
3	16952.216	23	16953.766	43	16955.540	63	16957.153
4	16952.304	24	16953.870	44	16955.635	64	16957.262
5	16952.422	25	16953.920	45	16955.683	65	16957.383
6	16952.466	26	16953.967	46	16955.756	66	16957.466
7	16952.514	27	16954.140	47	16955.810	67	16957.584
8	16952.617	28	16954.282	48	16955.910	68	16957.622
9	16952.735	29	16954.356	49	16956.006	69	16957.696
10	16952.865	30	16954.433	50	16956.111	70	16957.799
11	16952.970	31	16954.508	51	16956.250	71	16957.848
12	16953.031	32	16954.557	52	16956.309	72	16957.904
13	16953.077	33	16954.641	53	16956.340	73	16957.960
14	16953.203	34	16954.733	54	16956.412	74	16958.064
15	16953.247	35	16954.789	55	16956.511	75	16958.207
16	16953.291	36	16954.901	56	16956.635		
17	16953.339	37	16954.966	57	16956.671		
18	16953.386	38	16955.056	58	16956.745		
19	16953.459	39	16955.176	59	16956.830		
20	16953.534	40	16955.306	60	16956.895		

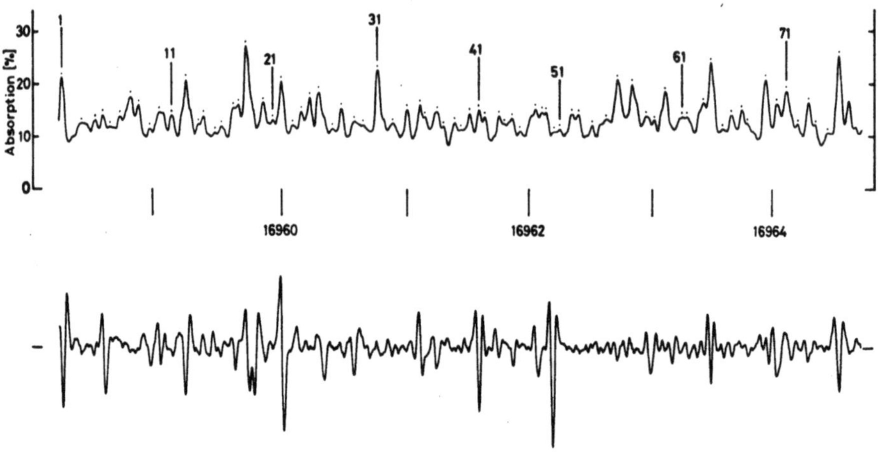

1	16958.297	21	16959.938	41	16961.584	61	16963.252
2	16958.436	22	16960.004	42	16961.631	62	16963.288
3	16958.553	23	16960.094	43	16961.747	63	16963.426
4	16958.610	24	16960.163	44	16961.793	64	16963.486
5	16958.673	25	16960.229	45	16961.852	65	16963.584
6	16958.748	26	16960.303	46	16961.921	66	16963.649
7	16958.832	27	16960.413	47	16962.048	67	16963.738
8	16958.891	28	16960.483	48	16962.100	68	16963.787
9	16958.977	29	16960.586	49	16962.134	69	16963.945
10	16959.052	30	16960.655	50	16962.192	70	16964.038
11	16959.146	31	16960.765	51	16962.243	71	16964.126
12	16959.257	32	16960.885	52	16962.343	72	16964.217
13	16959.358	33	16961.001	53	16962.396	73	16964.308
14	16959.398	34	16961.107	54	16962.512	74	16964.367
15	16959.488	35	16961.148	55	16962.625	75	16964.480
16	16959.542	36	16961.240	56	16962.724	76	16964.576
17	16959.638	37	16961.298	57	16962.844		
18	16959.674	38	16961.386	58	16962.960		
19	16959.732	39	16961.436	59	16963.020		
20	16959.863	40	16961.506	60	16963.114		

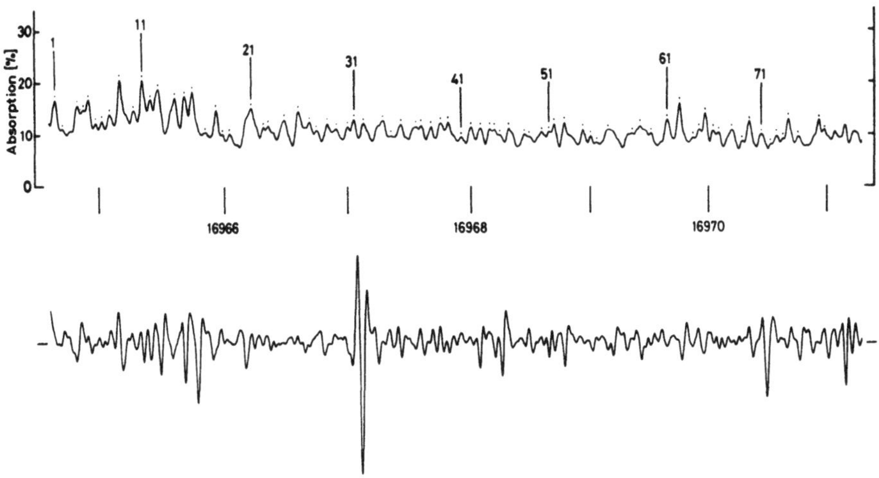

1	16964.662	21	16966.224	41	16967.925	61	16969.643
2	16964.721	22	16966.320	42	16968.006	62	16969.749
3	16964.838	23	16966.363	43	16968.083	63	16969.866
4	16964.882	24	16966.490	44	16968.160	64	16969.920
5	16964.924	25	16966.606	45	16968.204	65	16969.970
6	16964.978	26	16966.697	46	16968.248	66	16970.037
7	16965.029	27	16966.755	47	16968.319	67	16970.087
8	16965.088	28	16966.841	48	16968.447	68	16970.185
9	16965.172	29	16966.910	49	16968.490	69	16970.279
10	16965.280	30	16967.001	50	16968.593	70	16970.336
11	16965.353	31	16967.047	51	16968.656	71	16970.436
12	16965.415	32	16967.125	52	16968.703	72	16970.516
13	16965.479	33	16967.282	53	16968.789	73	16970.568
14	16965.611	34	16967.348	54	16968.946	74	16970.620
15	16965.688	35	16967.428	55	16969.010	75	16970.673
16	16965.747	36	16967.551	56	16969.058	76	16970.762
17	16965.849	37	16967.597	57	16969.159	77	16970.937
18	16965.937	38	16967.674	58	16969.347	78	16970.982
19	16965.989	39	16967.753	59	16969.419		
20	16966.048	40	16967.814	60	16969.511		

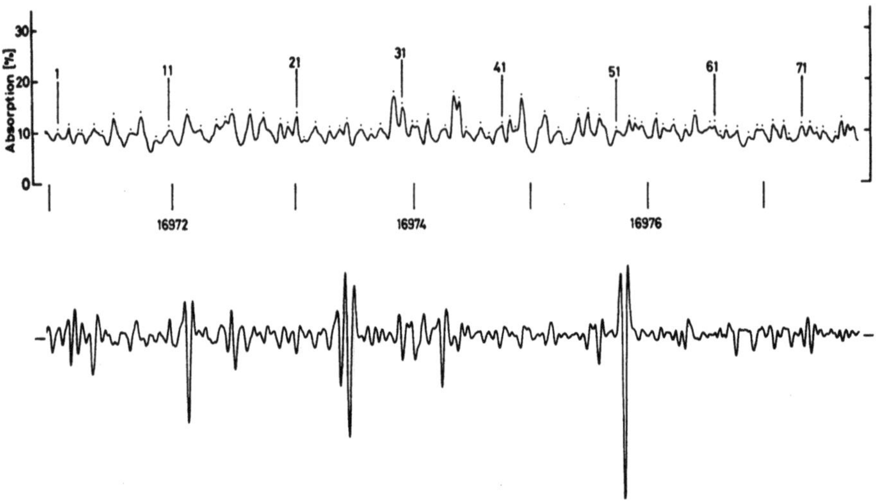

1	16971.075	21	16973.025	41	16974.749	61	16976.573
2	16971.163	22	16973.086	42	16974.819	62	16976.621
3	16971.237	23	16973.184	43	16974.866	63	16976.667
4	16971.264	24	16973.301	44	16974.920	64	16976.765
5	16971.367	25	16973.383	45	16975.118	65	16976.870
6	16971.436	26	16973.444	46	16975.237	66	16976.941
7	16971.529	27	16973.558	47	16975.308	67	16976.984
8	16971.671	28	16973.646	48	16975.406	68	16977.088
9	16971.750	29	16973.721	49	16975.486	69	16977.178
10	16971.871	30	16973.832	50	16975.585	70	16977.228
11	16971.978	31	16973.903	51	16975.731	71	16977.335
12	16972.127	32	16973.988	52	16975.845	72	16977.407
13	16972.242	33	16974.028	53	16975.894	73	16977.468
14	16972.373	34	16974.124	54	16975.950	74	16977.527
15	16972.448	35	16974.265	55	16976.066	75	16977.638
16	16972.499	36	16974.340	56	16976.129	76	16977.690
17	16972.647	37	16974.387	57	16976.215		
18	16972.754	38	16974.446	58	16976.318		
19	16972.892	39	16974.573	59	16976.402		
20	16972.952	40	16974.644	60	16976.531		

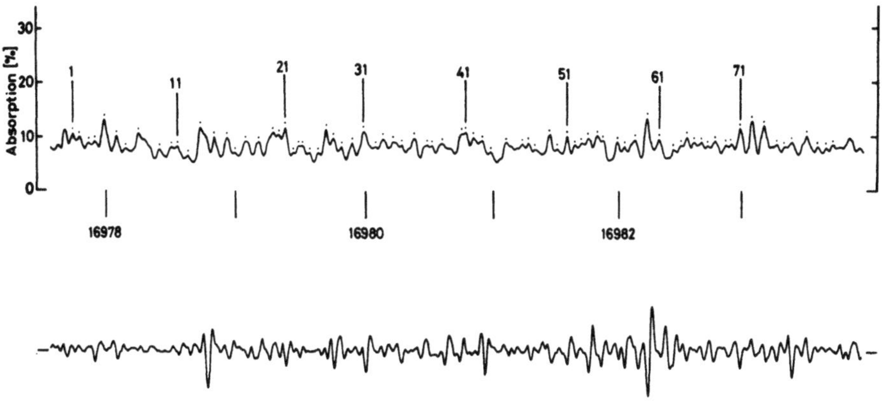

1	16977.747	21	16979.367	41	16980.763	61	16982.314
2	16977.795	22	16979.431	42	16980.826	62	16982.432
3	16977.862	23	16979.471	43	16980.895	63	16982.494
4	16977.912	24	16979.503	44	16980.965	64	16982.542
5	16977.983	25	16979.556	45	16981.083	65	16982.599
6	16978.073	26	16979.624	46	16981.211	66	16982.666
7	16978.140	27	16979.688	47	16981.262	67	16982.716
8	16978.242	28	16979.741	48	16981.313	68	16982.774
9	16978.398	29	16979.806	49	16981.433	69	16982.861
10	16978.490	30	16979.886	50	16981.496	70	16982.910
11	16978.539	31	16979.975	51	16981.573	71	16982.980
12	16978.619	32	16980.061	52	16981.636	72	16983.071
13	16978.712	33	16980.121	53	16981.693	73	16983.169
14	16978.821	34	16980.199	54	16981.751	74	16983.276
15	16978.920	35	16980.267	55	16981.820	75	16983.342
16	16978.980	36	16980.361	56	16981.867	76	16983.396
17	16979.062	37	16980.461	57	16981.981	77	16983.525
18	16979.159	38	16980.502	58	16982.040	78	16983.612
19	16979.273	39	16980.583	59	16982.127	79	16983.683
20	16979.315	40	16980.730	60	16982.222	80	16983.737

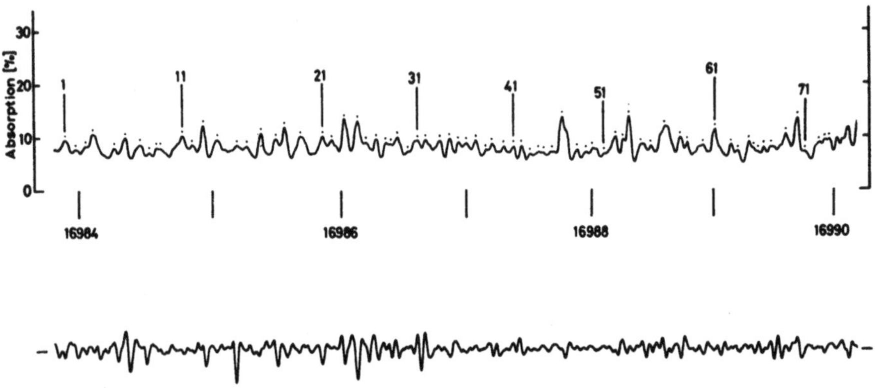

1	16983.876	21	16985.851	41	16987.378	61	16989.022
2	16983.955	22	16985.927	42	16987.440	62	16989.155
3	16984.027	23	16986.020	43	16987.505	63	16989.210
4	16984.082	24	16986.121	44	16987.573	64	16989.305
5	16984.245	25	16986.179	45	16987.614	65	16989.343
6	16984.328	26	16986.270	46	16987.684	66	16989.379
7	16984.440	27	16986.354	47	16987.764	67	16989.424
8	16984.511	28	16986.388	48	16987.887	68	16989.483
9	16984.561	29	16986.445	49	16987.958	69	16989.612
10	16984.597	30	16986.528	50	16988.014	70	16989.709
11	16984.762	31	16986.613	51	16988.105	71	16989.770
12	16984.839	32	16986.674	52	16988.205	72	16989.895
13	16984.922	33	16986.783	53	16988.272	73	16989.942
14	16985.031	34	16986.869	54	16988.317	74	16989.978
15	16985.190	35	16986.937	55	16988.408		
16	16985.264	36	16987.001	56	16988.493		
17	16985.375	37	16987.079	57	16988.611		
18	16985.486	38	16987.161	58	16988.735		
19	16985.559	39	16987.209	59	16988.796		
20	16985.688	40	16987.302	60	16988.924		

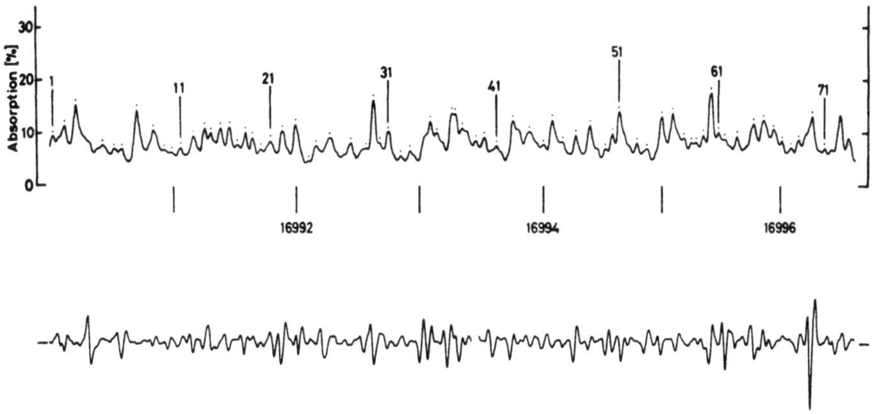

1	16990.047	21	16991.776	41	16993.613	61	16995.479
2	16990.137	22	16991.878	42	16993.749	62	16995.531
3	16990.231	23	16991.986	43	16993.887	63	16995.632
4	16990.439	24	16992.088	44	16993.999	64	16995.772
5	16990.534	25	16992.158	45	16994.074	65	16995.855
6	16990.586	26	16992.263	46	16994.159	66	16995.944
7	16990.714	27	16992.434	47	16994.271	67	16996.015
8	16990.844	28	16992.561	48	16994.391	68	16996.094
9	16990.928	29	16992.620	49	16994.524	69	16996.160
10	16990.993	30	16992.675	50	16994.579	70	16996.277
11	16991.059	31	16992.742	51	16994.641	71	16996.384
12	16991.161	32	16992.841	52	16994.790		
13	16991.251	33	16992.919	53	16994.875		
14	16991.300	34	16993.084	54	16995.001		
15	16991.379	35	16993.133	55	16995.091		
16	16991.447	36	16993.256	56	16995.190		
17	16991.511	37	16993.287	57	16995.252		
18	16991.578	38	16993.339	58	16995.292		
19	16991.637	39	16993.449	59	16995.355		
20	16991.703	40	16993.519	60	16995.421		

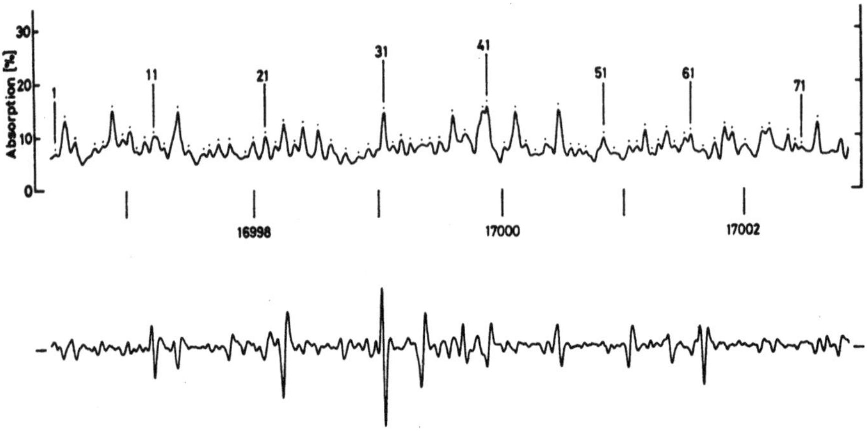

1	16996.445	21	16998.098	41	16999.885	61	17001.559
2	16996.521	22	16998.177	42	17000.029	62	17001.656
3	16996.596	23	16998.243	43	17000.118	63	17001.759
4	16996.748	24	16998.331	44	17000.199	64	17001.839
5	16996.811	25	16998.398	45	17000.277	65	17001.904
6	16996.896	26	16998.520	46	17000.359	66	17002.010
7	16996.972	27	16998.623	47	17000.466	67	17002.159
8	16997.031	28	16998.742	48	17000.569	68	17002.219
9	16997.082	29	16998.846	49	17000.638	69	17002.380
10	16997.142	30	16998.927	50	17000.693	70	17002.436
11	16997.215	31	16999.053	51	17000.838	71	17002.492
12	16997.292	32	16999.125	52	17000.923	72	17002.628
13	16997.405	33	16999.192	53	17001.048		
14	16997.483	34	16999.267	54	17001.115		
15	16997.605	35	16999.360	55	17001.181		
16	16997.653	36	16999.427	56	17001.237		
17	16997.726	37	16999.501	57	17001.289		
18	16997.816	38	16999.610	58	17001.360		
19	16997.937	39	16999.710	59	17001.427		
20	16997.996	40	16999.853	60	17001.509		

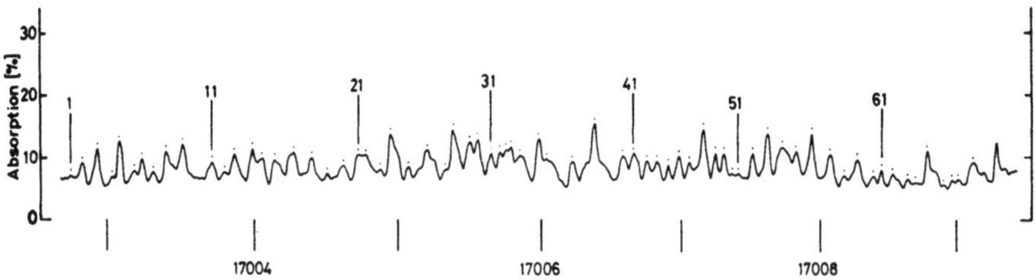

1	17002.751	21	17004.723	41	17006.669	61	17008.442
2	17002.831	22	17004.780	42	17006.761	62	17008.526
3	17002.937	23	17004.878	43	17006.833	63	17008.635
4	17003.034	24	17004.954	44	17006.914	64	17008.693
5	17003.088	25	17005.072	45	17006.987	65	17008.784
6	17003.186	26	17005.196	46	17007.063	66	17008.911
7	17003.241	27	17005.327	47	17007.170	67	17008.967
8	17003.317	28	17005.385	48	17007.252	68	17009.015
9	17003.408	29	17005.499	49	17007.311	69	17009.118
10	17003.523	30	17005.558	50	17007.368		
11	17003.722	31	17005.648	51	17007.414		
12	17003.801	32	17005.710	52	17007.518		
13	17003.870	33	17005.755	53	17007.625		
14	17003.991	34	17005.790	54	17007.726		
15	17004.060	35	17005.857	55	17007.831		
16	17004.144	36	17005.983	56	17007.948		
17	17004.276	37	17006.037	57	17008.080		
18	17004.403	38	17006.220	58	17008.176		
19	17004.511	39	17006.384	59	17008.267		
20	17004.624	40	17006.589	60	17008.386		

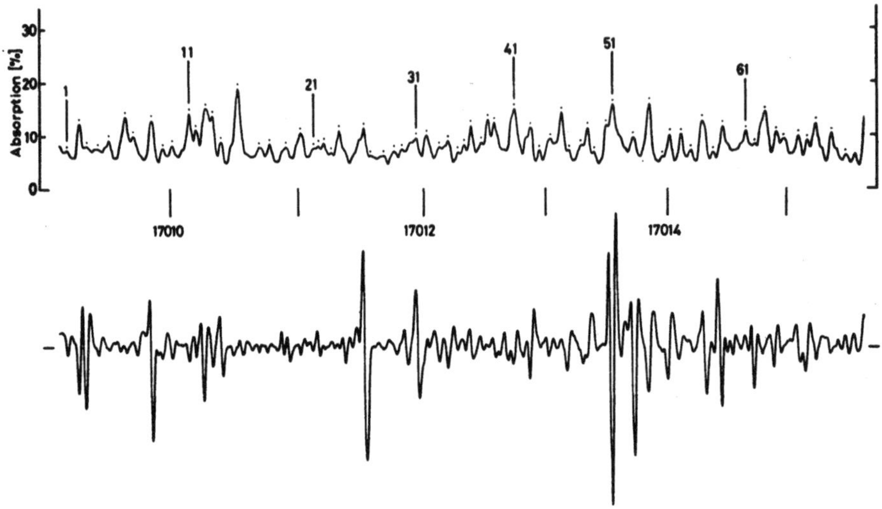

1	17009.202	21	17011.125	41	17012.736	61	17014.652
2	17009.298	22	17011.162	42	17012.871	62	17014.731
3	17009.359	23	17011.207	43	17012.946	63	17014.819
4	17009.431	24	17011.262	44	17013.039	64	17014.913
5	17009.530	25	17011.332	45	17013.129	65	17014.979
6	17009.662	26	17011.526	46	17013.191	66	17015.096
7	17009.720	27	17011.580	47	17013.285	67	17015.170
8	17009.863	28	17011.685	48	17013.344	68	17015.240
9	17009.948	29	17011.765	49	17013.397	69	17015.376
10	17010.020	30	17011.829	50	17013.497	70	17015.502
11	17010.151	31	17011.943	51	17013.551		
12	17010.202	32	17012.027	52	17013.718		
13	17010.280	33	17012.129	53	17013.855		
14	17010.333	34	17012.197	54	17013.958		
15	17010.402	35	17012.278	55	17014.018		
16	17010.539	36	17012.327	56	17014.115		
17	17010.701	37	17012.384	57	17014.195		
18	17010.782	38	17012.467	58	17014.290		
19	17010.909	39	17012.523	59	17014.381		
20	17011.023	40	17012.574	60	17014.456		

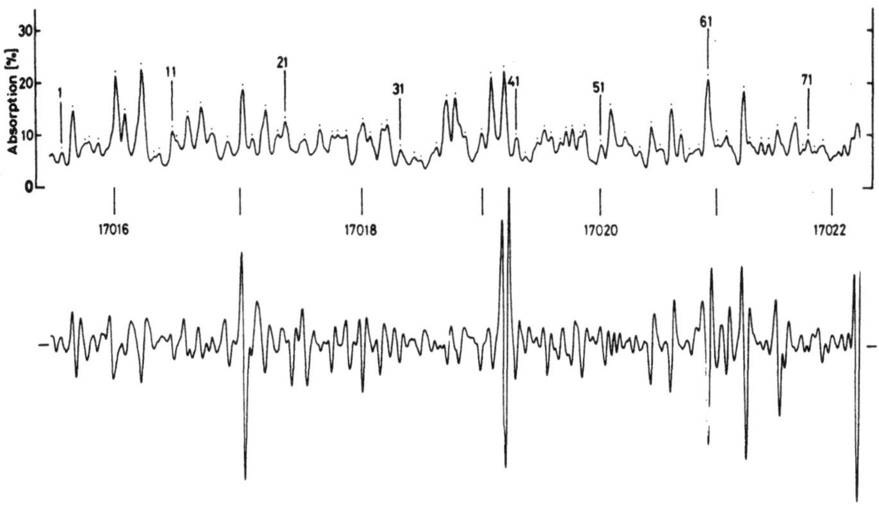

1	17015.583	21	17017.370	41	17019.279	61	17020.928
2	17015.670	22	17017.525	42	17019.354	62	17021.011
3	17015.758	23	17017.654	43	17019.460	63	17021.085
4	17015.802	24	17017.754	44	17019.521	64	17021.231
5	17015.872	25	17017.801	45	17019.580	65	17021.279
6	17016.016	26	17017.874	46	17019.660	66	17021.375
7	17016.085	27	17018.009	47	17019.707	67	17021.437
8	17016.221	28	17018.066	48	17019.759	68	17021.514
9	17016.318	29	17018.166	49	17019.820	69	17021.675
10	17016.356	30	17018.207	50	17019.861	70	17021.741
11	17016.466	31	17018.319	51	17020.008	71	17021.783
12	17016.504	32	17018.430	52	17020.092	72	17021.844
13	17016.589	33	17018.484	53	17020.206	73	17021.922
14	17016.696	34	17018.615	54	17020.331		
15	17016.775	35	17018.701	55	17020.436		
16	17016.901	36	17018.774	56	17020.507		
17	17017.025	37	17018.856	57	17020.610		
18	17017.101	38	17018.990	58	17020.692		
19	17017.210	39	17019.071	59	17020.769		
20	17017.314	40	17019.179	60	17020.848		

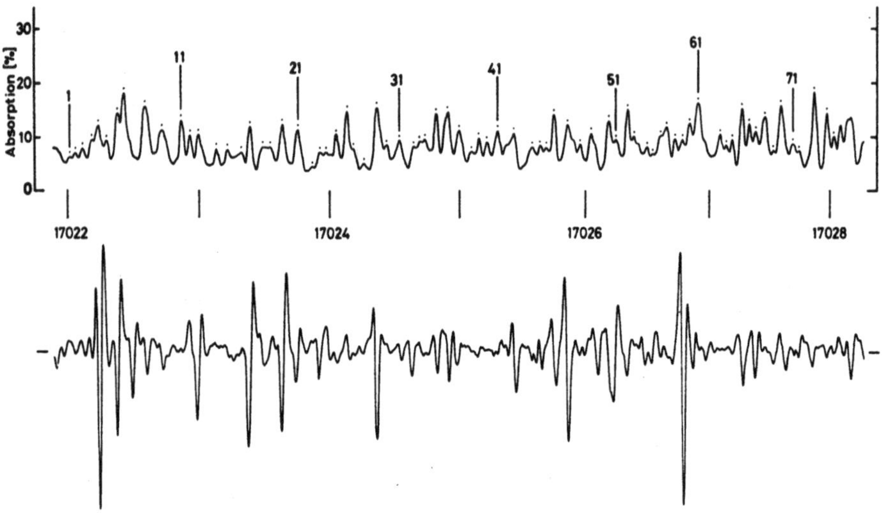

1	17022.015	21	17023.756	41	17025.305	61	17026.912
2	17022.057	22	17023.866	42	17025.430	62	17027.087
3	17022.111	23	17023.928	43	17025.578	63	17027.138
4	17022.180	24	17023.975	44	17025.644	64	17027.185
5	17022.227	25	17024.052	45	17025.691	65	17027.267
6	17022.292	26	17024.136	46	17025.753	66	17027.329
7	17022.382	27	17024.186	47	17025.860	67	17027.384
8	17022.433	28	17024.260	48	17025.959	68	17027.462
9	17022.591	29	17024.364	49	17026.047	69	17027.526
10	17022.718	30	17024.437	50	17026.187	70	17027.596
11	17022.866	31	17024.535	51	17026.247	71	17027.691
12	17022.933	32	17024.647	52	17026.343	72	17027.744
13	17023.000	33	17024.695	53	17026.392	73	17027.869
14	17023.136	34	17024.733	54	17026.495	74	17027.977
15	17023.224	35	17024.821	55	17026.537	75	17028.033
16	17023.326	36	17024.911	56	17026.600		
17	17023.395	37	17024.997	57	17026.652		
18	17023.488	38	17025.096	58	17026.720		
19	17023.544	39	17025.156	59	17026.781		
20	17023.638	40	17025.220	60	17026.841		

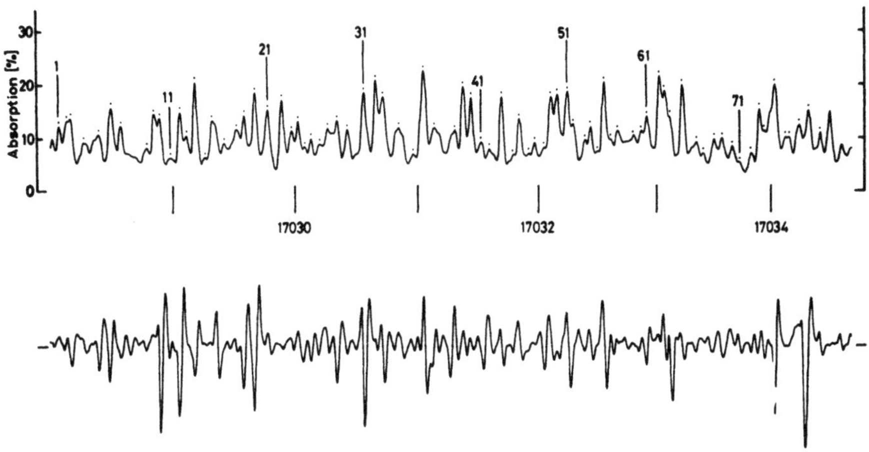

1	17028.088	21	17029.779	41	17031.519	61	17032.918
2	17028.150	22	17029.895	42	17031.586	62	17033.025
3	17028.178	23	17029.973	43	17031.690	63	17033.071
4	17028.281	24	17030.029	44	17031.787	64	17033.114
5	17028.396	25	17030.078	45	17031.844	65	17033.230
6	17028.499	26	17030.132	46	17031.921	66	17033.349
7	17028.577	27	17030.201	47	17031.965	67	17033.411
8	17028.784	28	17030.260	48	17032.045	68	17033.502
9	17028.845	29	17030.338	49	17032.105	69	17033.570
10	17028.892	30	17030.420	50	17032.156	70	17033.655
11	17028.978	31	17030.553	51	17032.241	71	17033.719
12	17029.059	32	17030.649	52	17032.289	72	17033.822
13	17029.107	33	17030.711	53	17032.390	73	17033.894
14	17029.181	34	17030.838	54	17032.439	74	17033.936
15	17029.255	35	17030.962	55	17032.496	75	17034.033
16	17029.321	36	17031.047	56	17032.553	76	17034.125
17	17029.418	37	17031.137	57	17032.611	77	17034.160
18	17029.518	38	17031.305	58	17032.667	78	17034.251
19	17029.585	39	17031.367	59	17032.805	79	17034.332
20	17029.674	40	17031.434	60	17032.869	80	17034.428

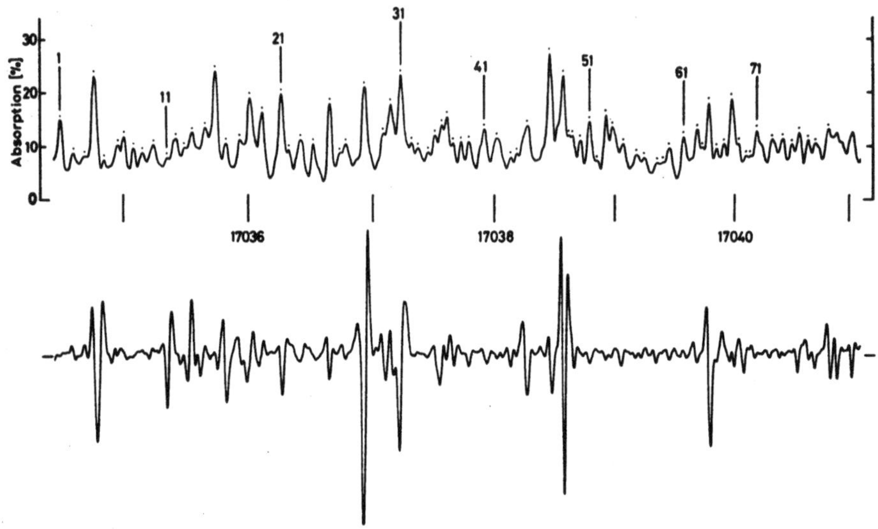

1	17034.509	21	17036.257	41	17037.915	61	17039.567
2	17034.607	22	17036.317	42	17038.013	62	17039.679
3	17034.686	23	17036.413	43	17038.127	63	17039.727
4	17034.773	24	17036.515	44	17038.173	64	17039.779
5	17034.849	25	17036.652	45	17038.264	65	17039.840
6	17034.960	26	17036.734	46	17038.445	66	17039.903
7	17035.009	27	17036.778	47	17038.559	67	17039.971
8	17035.083	28	17036.930	48	17038.603	68	17040.030
9	17035.151	29	17037.085	49	17038.641	69	17040.091
10	17035.236	30	17037.144	50	17038.701	70	17040.132
11	17035.348	31	17037.225	51	17038.779	71	17040.186
12	17035.420	32	17037.318	52	17038.844	72	17040.320
13	17035.479	33	17037.372	53	17038.918	73	17040.410
14	17035.548	34	17037.458	54	17038.974	74	17040.491
15	17035.651	35	17037.515	55	17039.064	75	17040.558
16	17035.736	36	17037.566	56	17039.172	76	17040.633
17	17035.819	37	17037.609	57	17039.239	77	17040.700
18	17035.934	38	17037.658	58	17039.347	78	17040.814
19	17036.014	39	17037.724	59	17039.394		
20	17036.108	40	17037.789	60	17039.447		

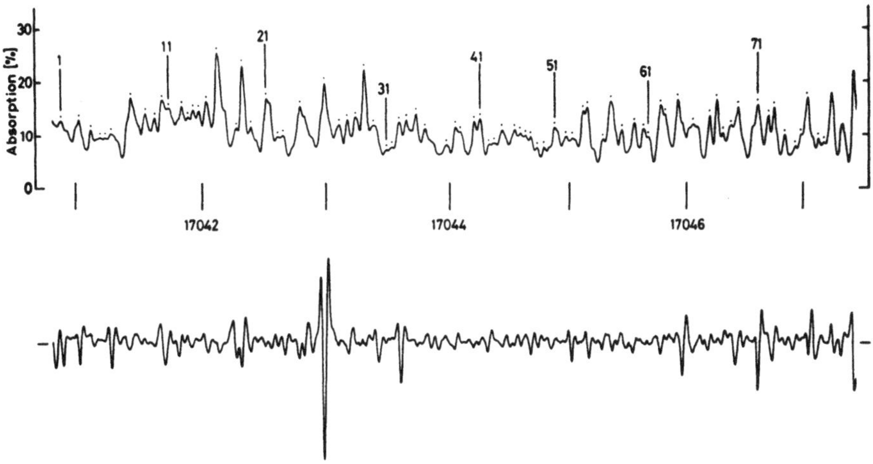

1	17040.885	21	17042.517	41	17044.255	61	17045.675
2	17041.030	22	17042.613	42	17044.329	62	17045.774
3	17041.125	23	17042.655	43	17044.368	63	17045.813
4	17041.198	24	17042.788	44	17044.435	64	17045.923
5	17041.230	25	17042.986	45	17044.547	65	17046.055
6	17041.278	26	17043.101	46	17044.596	66	17046.191
7	17041.444	27	17043.168	47	17044.638	67	17046.254
8	17041.553	28	17043.236	48	17044.682	68	17046.301
9	17041.627	29	17043.307	49	17044.733	69	17046.371
10	17041.691	30	17043.375	50	17044.794	70	17046.434
11	17041.738	31	17043.484	51	17044.883	71	17046.600
12	17041.841	32	17043.526	52	17044.977	72	17046.690
13	17041.891	33	17043.592	53	17045.031	73	17046.740
14	17041.931	34	17043.649	54	17045.126	74	17046.837
15	17041.973	35	17043.728	55	17045.157	75	17046.928
16	17042.031	36	17043.802	56	17045.288	76	17046.979
17	17042.121	37	17043.971	57	17045.357	77	17047.039
18	17042.272	38	17044.049	58	17045.445	78	17047.134
19	17042.322	39	17044.092	59	17045.552	79	17047.182
20	17042.391	40	17044.207	60	17045.629		

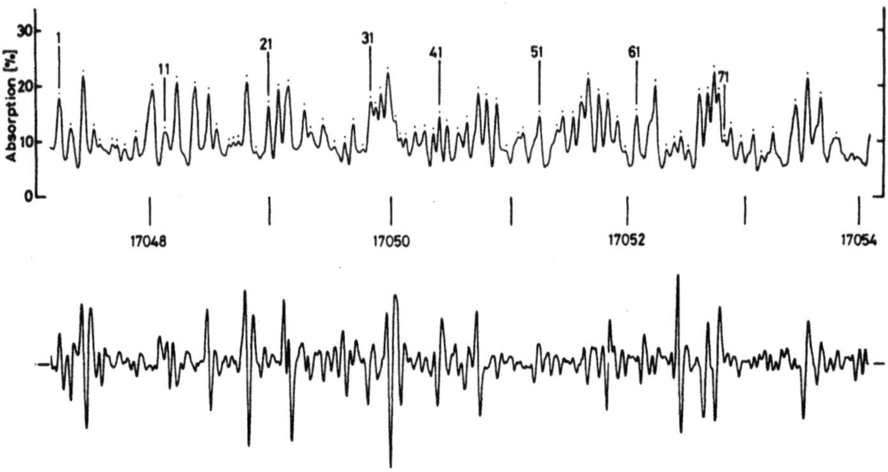

1	17047.252	21	17048.993	41	17050.409	61	17052.074
2	17047.348	22	17049.075	42	17050.473	62	17052.231
3	17047.450	23	17049.156	43	17050.566	63	17052.332
4	17047.535	24	17049.290	44	17050.639	64	17052.396
5	17047.594	25	17049.342	45	17050.733	65	17052.453
6	17047.695	26	17049.444	46	17050.803	66	17052.519
7	17047.736	27	17049.543	47	17050.889	67	17052.608
8	17047.798	28	17049.627	48	17050.968	68	17052.680
9	17047.888	29	17049.696	49	17051.070	69	17052.728
10	17048.025	30	17049.777	50	17051.110	70	17052.771
11	17048.130	31	17049.841	51	17051.249	71	17052.821
12	17048.227	32	17049.882	52	17051.401	72	17052.871
13	17048.378	33	17049.925	53	17051.446	73	17052.963
14	17048.493	34	17049.982	54	17051.535	74	17053.066
15	17048.563	35	17050.045	55	17051.603	75	17053.140
16	17048.668	36	17050.087	56	17051.660	76	17053.239
17	17048.700	37	17050.131	57	17051.752	77	17053.432
18	17048.749	38	17050.209	58	17051.826	78	17053.530
19	17048.820	39	17050.286	59	17051.911	79	17053.647
20	17048.893	40	17050.359	60	17051.982	80	17053.795

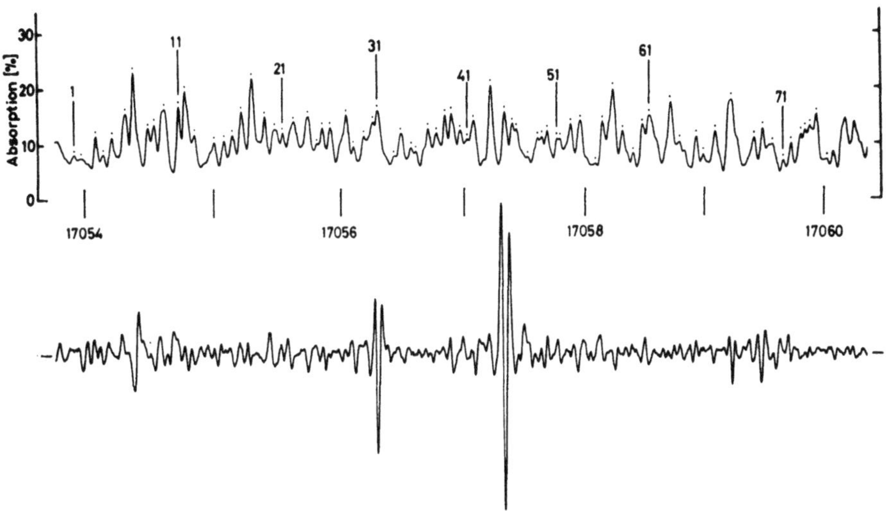

1	17053.920	21	17055.543	41	17057.021	61	17058.536
2	17053.982	22	17055.628	42	17057.071	62	17058.711
3	17054.084	23	17055.741	43	17057.208	63	17058.782
4	17054.137	24	17055.809	44	17057.322	64	17058.928
5	17054.205	25	17055.856	45	17057.384	65	17058.993
6	17054.307	26	17055.919	46	17057.418	66	17059.087
7	17054.374	27	17056.050	47	17057.511	67	17059.209
8	17054.487	28	17056.104	48	17057.593	68	17059.408
9	17054.539	29	17056.187	49	17057.627	69	17059.480
10	17054.621	30	17056.258	50	17057.678	70	17059.554
11	17054.731	31	17056.298	51	17057.763	71	17059.651
12	17054.786	32	17056.428	52	17057.789	72	17059.721
13	17054.856	33	17056.485	53	17057.876	73	17059.808
14	17055.006	34	17056.565	54	17057.961	74	17059.845
15	17055.077	35	17056.604	55	17058.082	75	17059.888
16	17055.143	36	17056.709	56	17058.144	76	17059.940
17	17055.215	37	17056.774	57	17058.232	77	17060.025
18	17055.297	38	17056.843	58	17058.308		
19	17055.397	39	17056.898	59	17058.395		
20	17055.480	40	17056.966	60	17058.475		

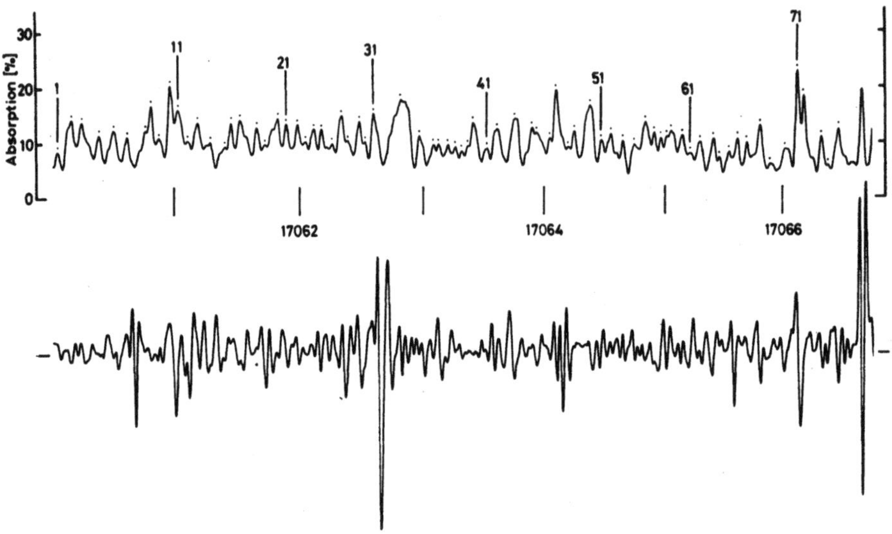

1	17060.088	21	17061.895	41	17063.528	61	17065.207
2	17060.194	22	17061.984	42	17063.609	62	17065.290
3	17060.276	23	17062.046	43	17063.755	63	17065.408
4	17060.412	24	17062.112	44	17063.900	64	17065.456
5	17060.531	25	17062.175	45	17063.937	65	17065.538
6	17060.638	26	17062.264	46	17064.046	66	17065.607
7	17060.782	27	17062.341	47	17064.095	67	17065.690
8	17060.825	28	17062.389	48	17064.195	68	17065.799
9	17060.889	29	17062.483	49	17064.247	69	17065.893
10	17060.977	30	17062.531	50	17064.377	70	17066.017
11	17061.036	31	17062.595	51	17064.471	71	17066.118
12	17061.119	32	17062.819	52	17064.546	72	17066.173
13	17061.191	33	17062.968	53	17064.649	73	17066.317
14	17061.292	34	17063.083	54	17064.742	74	17066.377
15	17061.410	35	17063.124	55	17064.841	75	17066.463
16	17061.457	36	17063.203	56	17064.913		
17	17061.529	37	17063.264	57	17064.968		
18	17061.661	38	17063.314	58	17065.014		
19	17061.727	39	17063.368	59	17065.051		
20	17061.827	40	17063.411	60	17065.152		

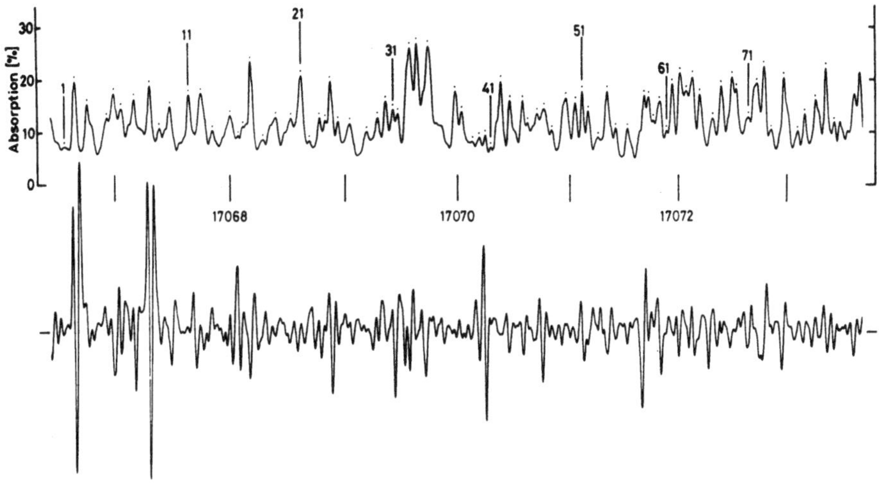

1	17066.580	21	17068.611	41	17070.290	61	17071.887
2	17066.668	22	17068.685	42	17070.379	62	17071.942
3	17066.771	23	17068.775	43	17070.457	63	17072.015
4	17066.936	24	17068.826	44	17070.572	64	17072.071
5	17066.992	25	17068.870	45	17070.612	65	17072.128
6	17067.055	26	17068.941	46	17070.697	66	17072.199
7	17067.165	27	17069.040	47	17070.761	67	17072.309
8	17067.301	28	17069.196	48	17070.833	68	17072.389
9	17067.386	29	17069.292	49	17070.959	69	17072.487
10	17067.474	30	17069.366	50	17071.046	70	17072.527
11	17067.644	31	17069.431	51	17071.108	71	17072.633
12	17087.748	32	17089.472	52	17071.162	72	17072.717
13	17067.846	33	17069.570	53	17071.255	73	17072.785
14	17068.002	34	17069.636	54	17071.339	74	17072.852
15	17068.069	35	17069.739	55	17071.419	75	17072.969
16	17068.113	36	17069.980	56	17071.530	76	17073.099
17	17068.178	37	17070.034	57	17071.681	77	17073.169
18	17068.282	38	17070.126	58	17071.723	78	17073.273
19	17068.396	39	17070.189	59	17071.764	79	17073.368
20	17068.527	40	17070.241	60	17071.826	80	17073.446

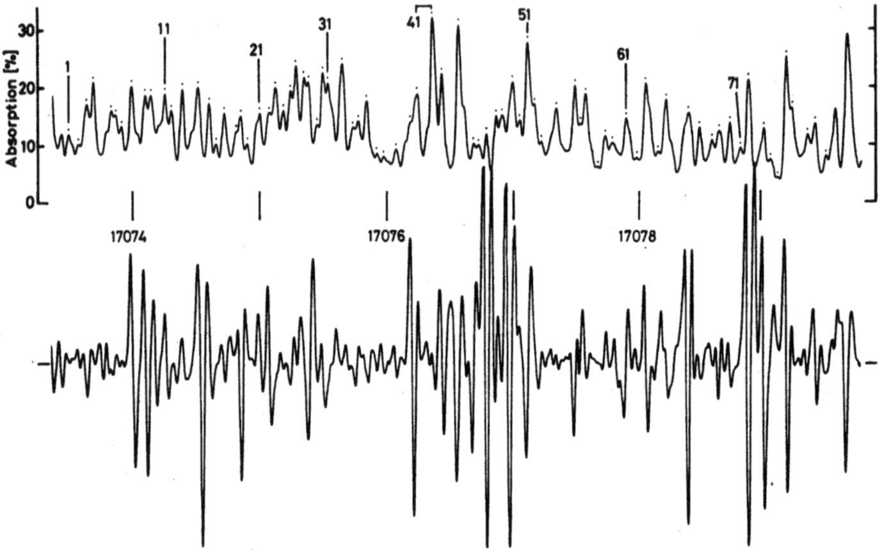

1	17073.502	21	17075.003	41	17076.356	61	17077.893
2	17073.580	22	17075.078	42	17076.433	62	17077.984
3	17073.642	23	17075.125	43	17076.563	63	17078.057
4	17073.694	24	17075.187	44	17076.690	64	17078.153
5	17073.839	25	17075.241	45	17076.726	65	17078.218
6	17073.873	26	17075.284	46	17076.790	66	17078.405
7	17073.920	27	17075.349	47	17076.864	67	17078.495
8	17073.997	28	17075.381	48	17076.913	68	17078.595
9	17074.103	29	17075.453	49	17076.997	69	17078.656
10	17074.145	30	17075.497	50	17077.050	70	17078.745
11	17074.257	31	17075.537	51	17077.115	71	17078.832
12	17074.308	32	17075.651	52	17077.169	72	17078.903
13	17074.393	33	17075.736	53	17077.225	73	17079.027
14	17074.514	34	17075.776	54	17077.341	74	17079.087
15	17074.603	35	17075.842	55	17077.491	75	17079.147
16	17074.660	36	17075.924	56	17077.532	76	17079.220
17	17074.724	37	17075.976	57	17077.576	77	17079.267
18	17074.819	38	17076.072	58	17077.670	78	17079.398
19	17074.854	39	17076.185	59	17077.729	79	17079.456
20	17074.906	40	17076.237	60	17077.784	80	17079.551

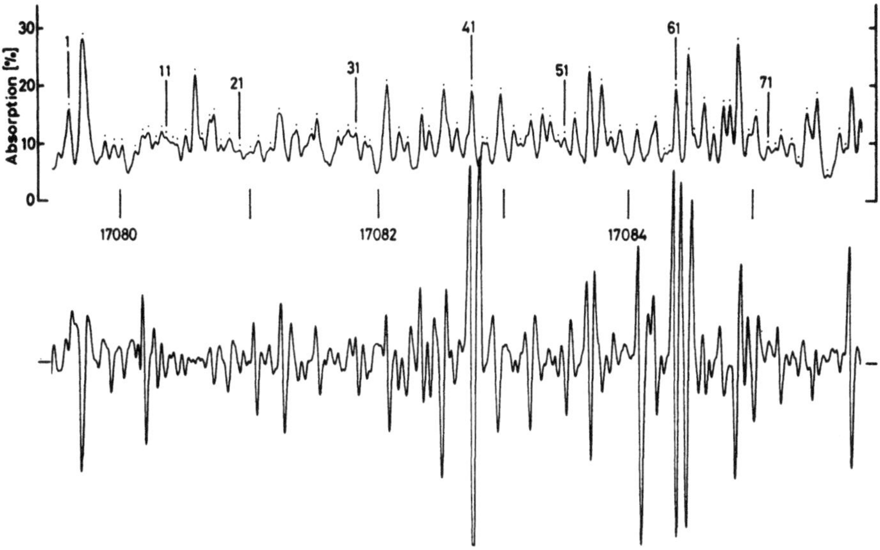

1	17079.634	21	17080.926	41	17082.746	61	17084.378
2	17079.738	22	17081.005	42	17082.837	62	17084.480
3	17079.897	23	17081.070	43	17082.870	63	17084.530
4	17079.962	24	17081.164	44	17082.979	64	17084.608
5	17080.023	25	17081.236	45	17083.080	65	17084.686
6	17080.123	26	17081.364	46	17083.137	66	17084.763
7	17080.178	27	17081.478	47	17083.220	67	17084.815
8	17080.220	28	17081.528	48	17083.313	68	17084.883
9	17080.276	29	17081.690	49	17083.376	69	17084.973
10	17080.328	30	17081.763	50	17083.433	70	17085.026
11	17080.366	31	17081.824	51	17083.485	71	17085.126
12	17080.415	32	17081.896	52	17083.570	72	17085.184
13	17080.448	33	17081.934	53	17083.690	73	17085.231
14	17080.514	34	17082.068	54	17083.788	74	17085.322
15	17080.591	35	17082.162	55	17083.857	75	17085.386
16	17080.641	36	17082.227	56	17083.929	76	17085.449
17	17080.699	37	17082.340	57	17084.067	77	17085.530
18	17080.730	38	17082.402	58	17084.209	78	17085.617
19	17080.787	39	17082.517	59	17084.283	79	17085.721
20	17080.851	40	17082.621	60	17084.327	80	17085.776

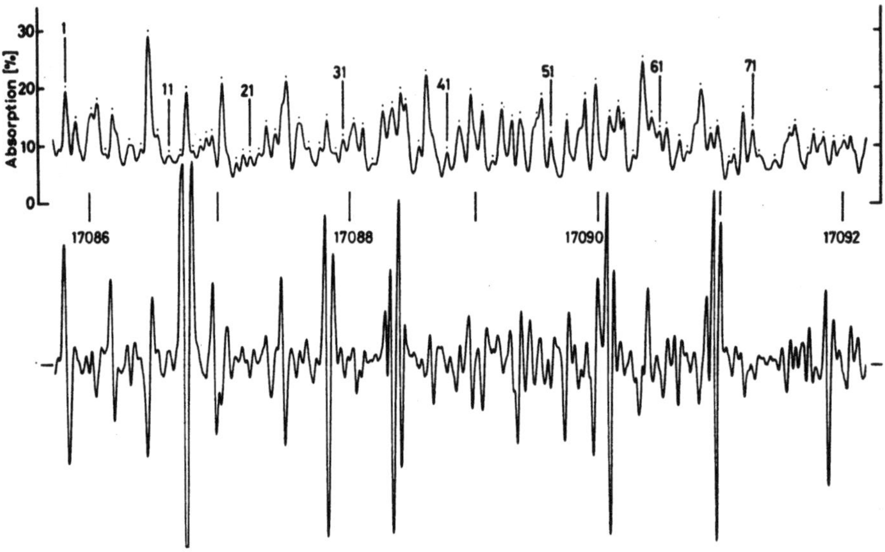

1	17085.825	21	17087.244	41	17088.770	61	17090.501
2	17085.902	22	17087.309	42	17088.870	62	17090.559
3	17086.019	23	17087.365	43	17088.964	63	17090.664
4	17086.061	24	17087.433	44	17089.056	64	17090.729
5	17086.128	25	17087.519	45	17089.137	65	17090.837
6	17086.185	26	17087.617	46	17089.212	66	17090.916
7	17086.316	27	17087.690	47	17089.290	67	17090.970
8	17086.404	28	17087.775	48	17089.357	68	17091.065
9	17086.462	29	17087.827	49	17089.494	69	17091.104
10	17086.534	30	17087.898	50	17089.532	70	17091.180
11	17086.619	31	17087.955	51	17089.601	71	17091.257
12	17086.706	32	17088.039	52	17089.737	72	17091.310
13	17086.755	33	17088.107	53	17089.890	73	17091.437
14	17086.802	34	17088.184	54	17089.978	74	17091.555
15	17086.861	35	17088.262	55	17090.096	75	17091.597
16	17086.906	36	17088.332	56	17090.165	76	17091.708
17	17086.952	37	17088.396	57	17090.211	77	17091.751
18	17087.031	38	17088.436	58	17090.295	78	17091.834
19	17087.141	39	17088.545	59	17090.362	79	17091.919
20	17087.192	40	17088.603	60	17090.439	80	17092.007

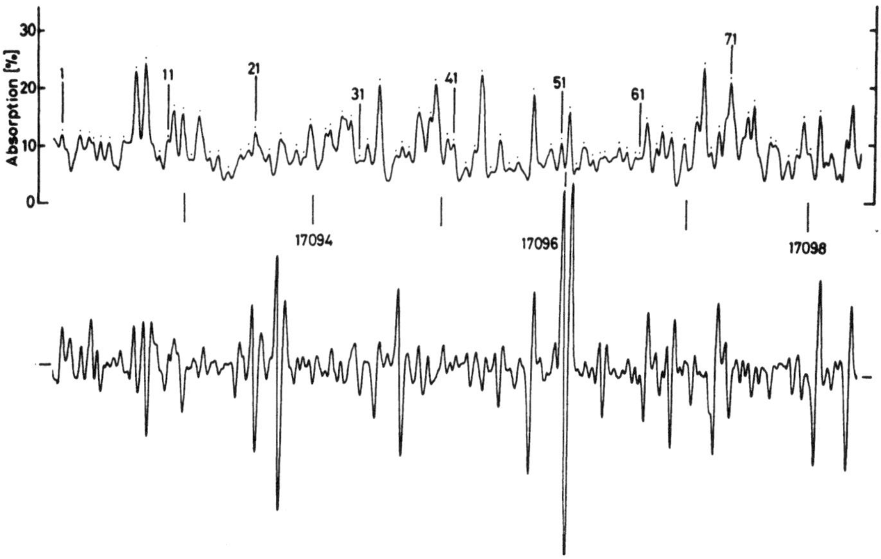

1	17092.066	21	17093.544	41	17095.092	61	17096.609
2	17092.205	22	17093.650	42	17095.194	62	17096.665
3	17092.271	23	17093.743	43	17095.263	63	17096.742
4	17092.304	24	17093.865	44	17095.319	64	17096.789
5	17092.362	25	17093.923	45	17095.391	65	17096.864
6	17092.427	26	17093.973	46	17095.465	66	17096.973
7	17092.540	27	17094.095	47	17095.537	67	17097.077
8	17092.632	28	17094.131	48	17095.610	68	17097.132
9	17092.711	29	17094.222	49	17095.736	69	17097.182
10	17092.815	30	17094.292	50	17095.879	70	17097.252
11	17092.877	31	17094.361	51	17095.965	71	17097.344
12	17092.923	32	17094.422	52	17096.032	72	17097.484
13	17092.992	33	17094.516	53	17096.092	73	17097.535
14	17093.057	34	17094.651	54	17096.152	74	17097.671
15	17093.116	35	17094.694	55	17096.225	75	17097.710
16	17093.196	36	17094.746	56	17096.288	76	17097.819
17	17093.257	37	17094.822	57	17096.321	77	17097.894
18	17093.334	38	17094.913	58	17096.438	78	17097.954
19	17093.426	39	17094.956	59	17096.500	79	17098.004
20	17093.490	40	17095.045	60	17096.573	80	17098.083

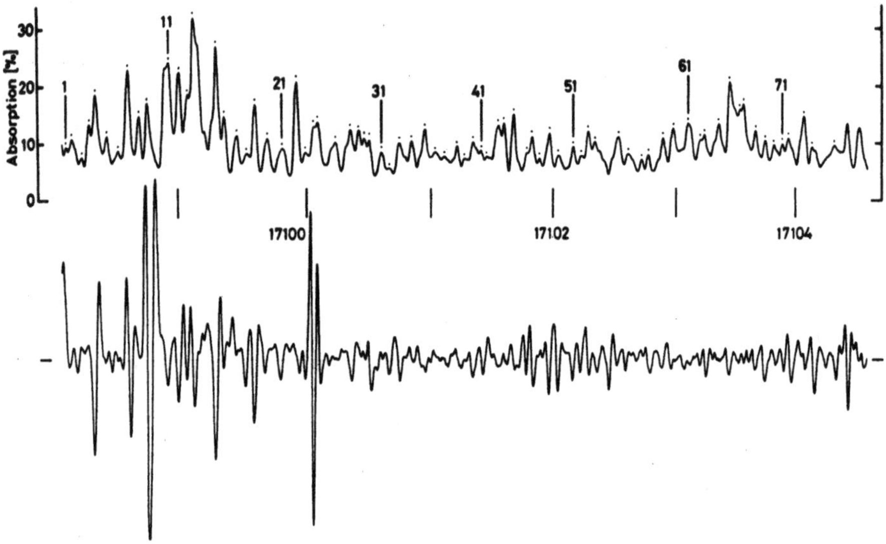

1	17098.135	21	17099.810	41	17101.416	61	17103.105
2	17098.179	22	17099.923	42	17101.462	62	17103.202
3	17098.252	23	17099.990	43	17101.556	63	17103.236
4	17098.310	24	17100.057	44	17101.603	64	17103.357
5	17098.357	25	17100.091	45	17101.683	65	17103.452
6	17098.450	26	17100.233	46	17101.794	66	17103.535
7	17098.536	27	17100.356	47	17101.832	67	17103.566
8	17098.614	28	17100.426	48	17101.894	68	17103.671
9	17098.698	29	17100.477	49	17101.974	69	17103.755
10	17098.765	30	17100.514	50	17102.047	70	17103.828
11	17098.925	31	17100.610	51	17102.168	71	17103.900
12	17099.009	32	17100.669	52	17102.236	72	17103.954
13	17099.076	33	17100.746	53	17102.295	73	17104.083
14	17099.125	34	17100.841	54	17102.345	74	17104.152
15	17099.303	35	17100.951	55	17102.533	75	17104.339
16	17099.362	36	17101.037	56	17102.615		
17	17099.461	37	17101.111	57	17102.727		
18	17099.539	38	17101.215	58	17102.783		
19	17099.603	39	17101.275	59	17102.902		
20	17099.696	40	17101.348	60	17102.980		

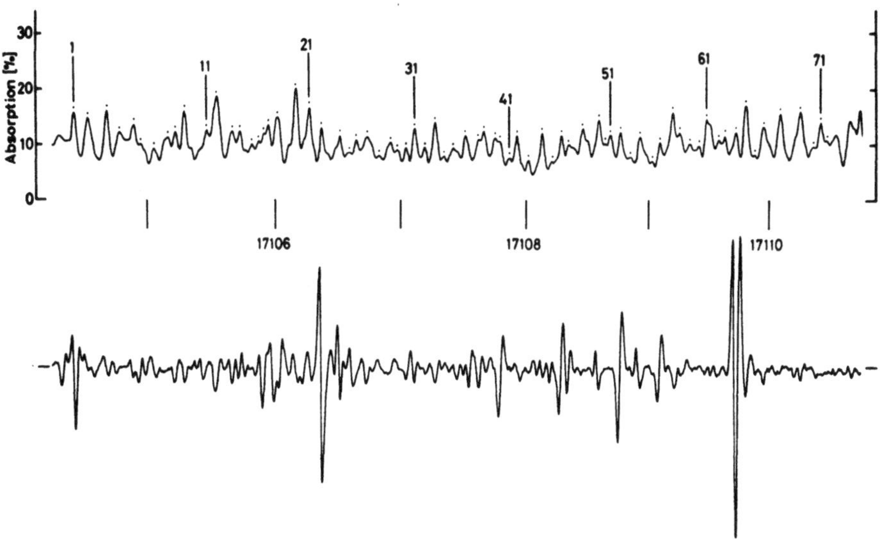

1	17104.446	21	17106.277	41	17107.869	61	17109.483
2	17104.549	22	17106.373	42	17107.924	62	17109.583
3	17104.690	23	17106.524	43	17108.020	63	17109.633
4	17104.786	24	17106.596	44	17108.132	64	17109.728
5	17104.890	25	17106.650	45	17108.207	65	17109.809
6	17104.951	26	17106.733	46	17108.287	66	17109.880
7	17105.050	27	17106.825	47	17108.348	67	17109.952
8	17105.159	28	17106.914	48	17108.464	68	17110.082
9	17105.215	29	17106.971	49	17108.596	69	17110.267
10	17105.287	30	17107.041	50	17108.647	70	17110.370
11	17105.462	31	17107.106	51	17108.690	71	17110.434
12	17105.542	32	17107.187	52	17108.769	72	17110.484
13	17105.661	33	17107.269	53	17108.856		
14	17105.726	34	17107.338	54	17108.928		
15	17105.819	35	17107.416	55	17109.030		
16	17105.873	36	17107.517	56	17109.088		
17	17105.914	37	17107.625	57	17109.195		
18	17105.949	38	17107.661	58	17109.251		
19	17106.021	39	17107.757	59	17109.337		
20	17106.172	40	17107.797	60	17109.417		

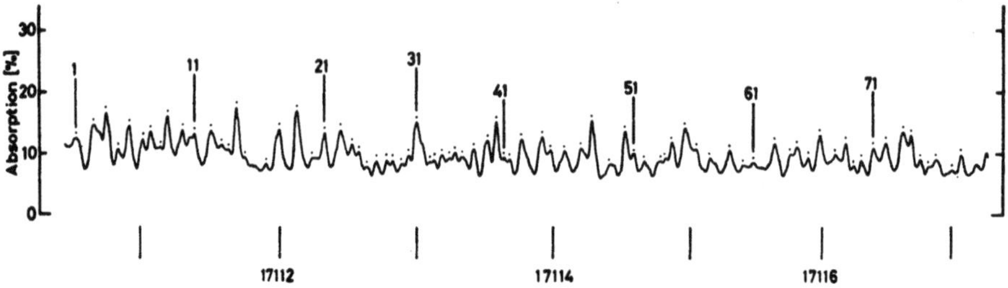

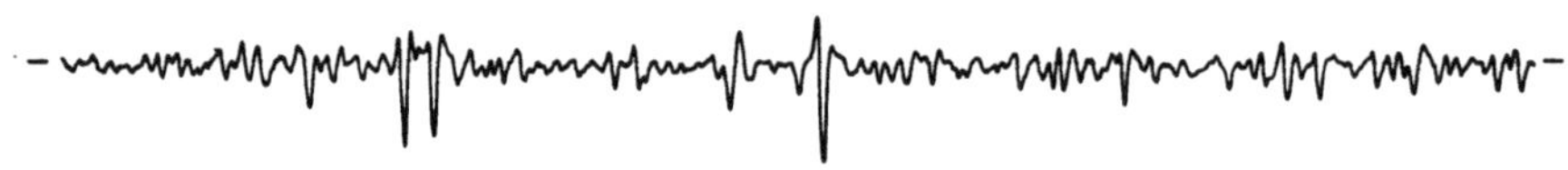

1	17110.560	21	17112.327	41	17113.636	61	17115.483
2	17110.682	22	17112.445	42	17113.677	62	17115.647
3	17110.765	23	17112.528	43	17113.763	63	17115.769
4	17110.846	24	17112.576	44	17113.915	64	17115.814
5	17110.924	25	17112.635	45	17113.974	65	17115.896
6	17111.019	26	17112.706	46	17114.078	66	17115.985
7	17111.082	27	17112.781	47	17114.200	67	17116.105
8	17111.141	28	17112.824	48	17114.278	68	17116.180
9	17111.200	29	17112.889	49	17114.408	69	17116.239
10	17111.305	30	17112.941	50	17114.523	70	17116.301
11	17111.389	31	17113.000	51	17114.586	71	17116.392
12	17111.504	32	17113.097	52	17114.666	72	17116.490
13	17111.576	33	17113.128	53	17114.788	73	17116.624
14	17111.628	34	17113.184	54	17114.822	74	17116.686
15	17111.688	35	17113.239	55	17114.872	75	17116.758
16	17111.752	36	17113.278	56	17114.969	76	17116.819
17	17111.904	37	17113.329	57	17115.055	77	17116.879
18	17112.001	38	17113.414	58	17115.157	78	17117.011
19	17112.128	39	17113.516	59	17115.305	79	17117.076
20	17112.234	40	17113.579	60	17115.406		

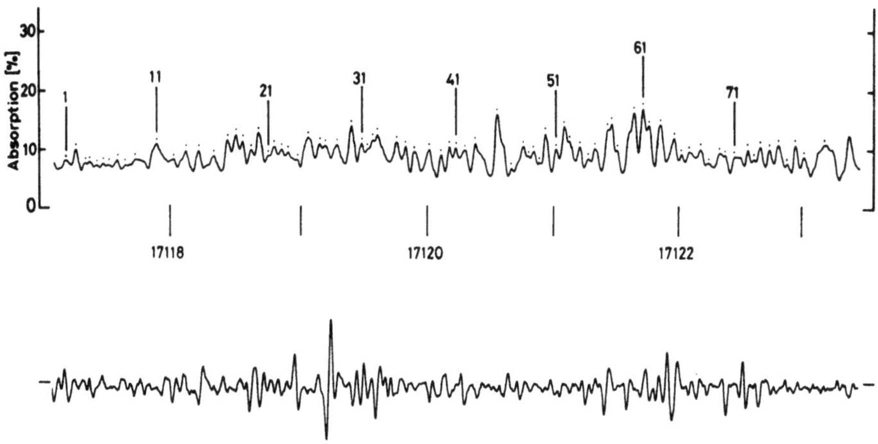

1	17117.194	21	17118.751	41	17120.226	61	17121.715
2	17117.271	22	17118.802	42	17120.294	62	17121.761
3	17117.338	23	17118.858	43	17120.383	63	17121.851
4	17117.382	24	17118.909	44	17120.551	64	17121.964
5	17117.432	25	17118.977	45	17120.665	65	17122.025
6	17117.481	26	17119.067	46	17120.752	66	17122.090
7	17117.523	27	17119.156	47	17120.795	67	17122.179
8	17117.593	28	17119.199	48	17120.838	68	17122.239
9	17117.667	29	17119.288	49	17120.885	69	17122.334
10	17117.731	30	17119.403	50	17120.935	70	17122.380
11	17117.899	31	17119.483	51	17121.022	71	17122.454
12	17118.031	32	17119.529	52	17121.091	72	17122.552
13	17118.122	33	17119.568	53	17121.133	73	17122.596
14	17118.217	34	17119.612	54	17121.214	74	17122.651
15	17118.326	35	17119.761	55	17121.286	75	17122.727
16	17118.434	36	17119.834	56	17121.336	76	17122.803
17	17118.496	37	17119.908	57	17121.434	77	17122.879
18	17118.548	38	17120.021	58	17121.463	78	17122.950
19	17118.616	39	17120.114	59	17121.510	79	17123.021
20	17118.678	40	17120.177	60	17121.645	80	17123.195

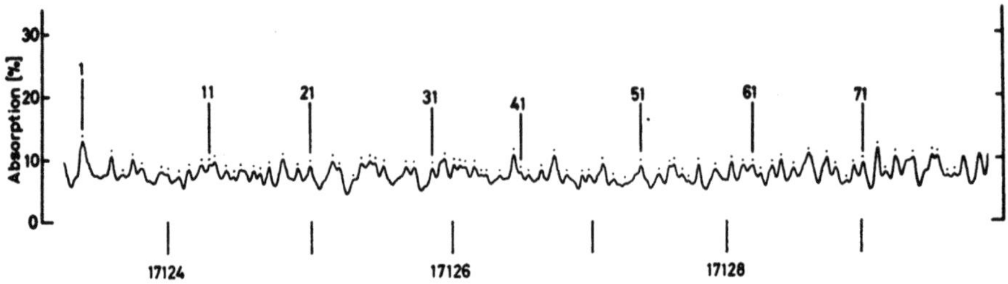

1	17123.400	21	17124.992	41	17126.484	61	17128.198
2	17123.593	22	17125.151	42	17126.541	62	17128.255
3	17123.663	23	17125.197	43	17126.632	63	17128.344
4	17123.747	24	17125.298	44	17126.726	64	17128.412
5	17123.810	25	17125.359	45	17126.815	65	17128.498
6	17123.956	26	17125.420	46	17126.926	66	17128.609
7	17123.997	27	17125.455	47	17126.975	67	17128.740
8	17124.076	28	17125.515	48	17127.076	68	17128.805
9	17124.143	29	17125.674	49	17127.156	69	17128.889
10	17124.229	30	17125.730	50	17127.246	70	17128.945
11	17124.283	31	17125.857	51	17127.358	71	17129.011
12	17124.320	32	17125.943	52	17127.494	72	17129.119
13	17124.402	33	17126.009	53	17127.573	73	17129.176
14	17124.455	34	17126.050	54	17127.606	74	17129.249
15	17124.502	35	17126.086	55	17127.666	75	17129.384
16	17124.599	36	17126.151	56	17127.715	76	17129.517
17	17124.645	37	17126.202	57	17127.788	77	17129.559
18	17124.710	38	17126.237	58	17127.914	78	17129.638
19	17124.808	39	17126.335	59	17128.040	79	17129.688
20	17124.908	40	17126.436	60	17128.125		

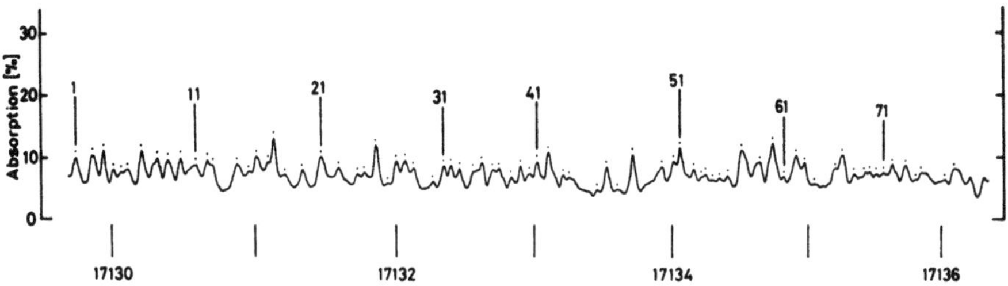

1	17129.754	21	17131.464	41	17133.022	61	17134.822
2	17129.874	22	17131.585	42	17133.105	62	17134.910
3	17129.949	23	17131.722	43	17133.205	63	17134.978
4	17130.017	24	17131.775	44	17133.245	64	17135.057
5	17130.068	25	17131.859	45	17133.445	65	17135.205
6	17130.109	26	17131.939	46	17133.516	66	17135.253
7	17130.211	27	17132.004	47	17133.593	67	17135.339
8	17130.320	28	17132.065	48	17133.709	68	17135.426
9	17130.391	29	17132.130	49	17133.922	69	17135.461
10	17130.484	30	17132.264	50	17134.012	70	17135.502
11	17130.593	31	17132.347	51	17134.060	71	17135.552
12	17130.672	32	17132.399	52	17134.161	72	17135.623
13	17130.707	33	17132.460	53	17134.213	73	17135.726
14	17130.877	34	17132.565	54	17134.249	74	17135.797
15	17130.958	35	17132.622	55	17134.299	75	17135.849
16	17131.016	36	17132.699	56	17134.344	76	17136.029
17	17131.092	37	17132.745	57	17134.407	77	17136.096
18	17131.132	38	17132.836	58	17134.513		
19	17131.203	39	17132.902	59	17134.642		
20	17131.334	40	17132.963	60	17134.743		

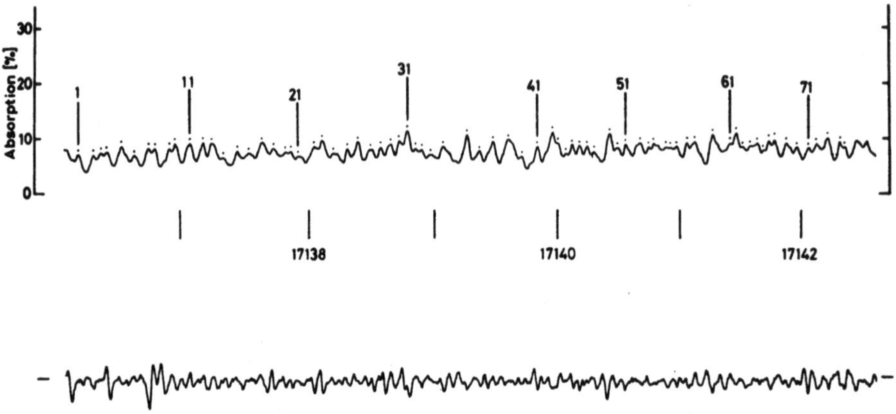

1	17136.223	21	17137.937	41	17139.836	61	17141.412
2	17136.334	22	17138.061	42	17139.958	62	17141.463
3	17136.390	23	17138.125	43	17140.005	63	17141.520
4	17136.434	24	17138.217	44	17140.069	64	17141.582
5	17136.548	25	17138.324	45	17140.119	65	17141.645
6	17136.652	26	17138.408	46	17140.179	66	17141.691
7	17136.763	27	17138.507	47	17140.240	67	17141.737
8	17136.811	28	17138.588	48	17140.299	68	17141.783
9	17136.922	29	17138.668	49	17140.425	69	17141.883
10	17136.966	30	17138.733	50	17140.489	70	17141.970
11	17137.083	31	17138.797	51	17140.554	71	17142.069
12	17137.185	32	17138.860	52	17140.677	72	17142.137
13	17137.253	33	17138.911	53	17140.744	73	17142.228
14	17137.347	34	17138.980	54	17140.785	74	17142.331
15	17137.460	35	17139.081	55	17140.890	75	17142.367
16	17137.548	36	17139.271	56	17140.924		
17	17137.650	37	17139.371	57	17140.972		
18	17137.742	38	17139.483	58	17141.060		
19	17137.835	39	17139.603	59	17141.119		
20	17137.880	40	17139.709	60	17141.274		

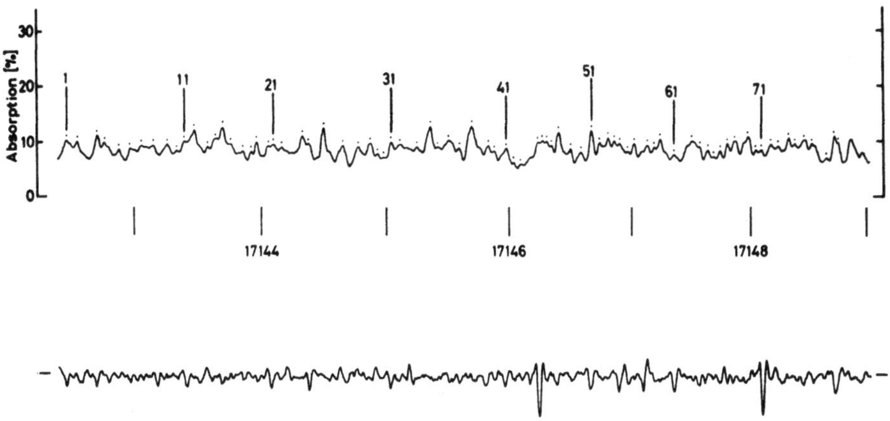

1	17142.464	21	17144.096	41	17145.979	61	17147.351
2	17142.555	22	17144.162	42	17146.046	62	17147.505
3	17142.711	23	17144.328	43	17146.095	63	17147.563
4	17142.764	24	17144.376	44	17146.239	64	17147.633
5	17142.876	25	17144.491	45	17146.275	65	17147.689
6	17142.969	26	17144.648	46	17146.308	66	17147.737
7	17143.054	27	17144.769	47	17146.345	67	17147.800
8	17143.139	28	17144.866	48	17146.404	68	17147.888
9	17143.256	29	17144.923	49	17146.497	69	17147.978
10	17143.331	30	17144.974	50	17146.580	70	17148.046
11	17143.387	31	17145.032	51	17146.670	71	17148.090
12	17143.468	32	17145.108	52	17146.737	72	17148.166
13	17143.575	33	17145.238	53	17146.809	73	17148.231
14	17143.641	34	17145.349	54	17146.857	74	17148.317
15	17143.690	35	17145.422	55	17146.901	75	17148.368
16	17143.759	36	17145.507	56	17146.970	76	17148.451
17	17143.854	37	17145.590	57	17147.021	77	17148.512
18	17143.920	38	17145.692	58	17147.128	78	17148.636
19	17143.965	39	17145.826	59	17147.191	79	17148.703
20	17144.056	40	17145.883	60	17147.236	80	17148.739

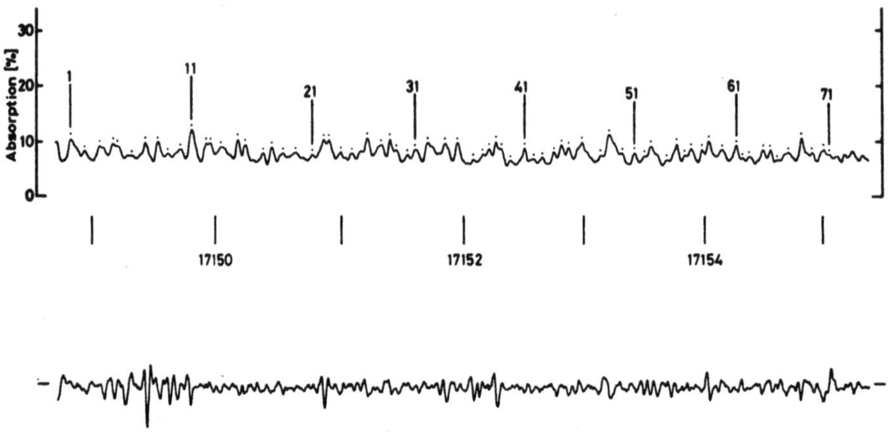

1	17148.849	21	17150.774	41	17152.503	61	17154.264
2	17148.951	22	17150.869	42	17152.577	62	17154.316
3	17149.069	23	17150.908	43	17152.652	63	17154.383
4	17149.177	24	17150.997	44	17152.748	64	17154.488
5	17149.216	25	17151.091	45	17152.815	65	17154.554
6	17149.327	26	17151.214	46	17152.875	66	17154.618
7	17149.435	27	17151.328	47	17152.991	67	17154.708
8	17149.536	28	17151.396	48	17153.147	68	17154.824
9	17149.615	29	17151.447	49	17153.217	69	17154.916
10	17149.722	30	17151.533	50	17153.319	70	17155.014
11	17149.814	31	17151.608	51	17153.427	71	17155.052
12	17149.926	32	17151.710	52	17153.509		
13	17149.964	33	17151.852	53	17153.563		
14	17150.056	34	17151.954	54	17153.694		
15	17150.185	35	17152.083	55	17153.776		
16	17150.243	36	17152.161	56	17153.841		
17	17150.383	37	17152.216	57	17153.893		
18	17150.454	38	17152.269	58	17153.983		
19	17150.540	39	17152.314	59	17154.040		
20	17150.644	40	17152.385	60	17154.149		

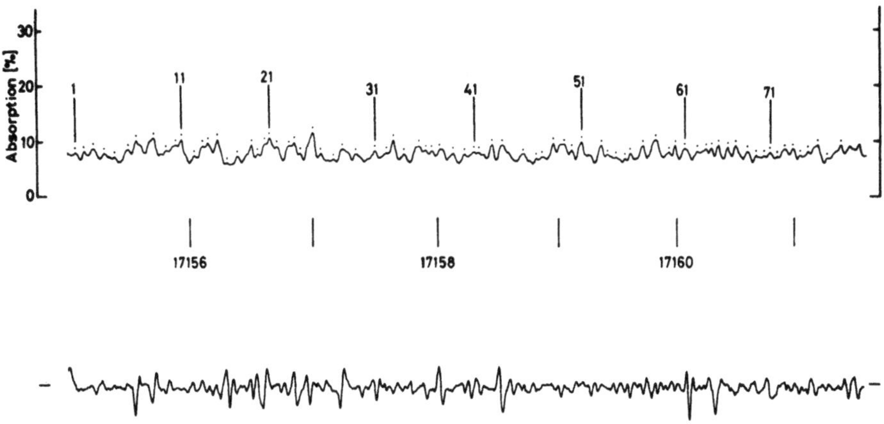

1	17155.121	21	17156.654	41	17158.296	61	17160.059
2	17155.179	22	17156.719	42	17158.345	62	17160.175
3	17155.248	23	17156.809	43	17158.442	63	17160.250
4	17155.334	24	17156.852	44	17158.539	64	17160.294
5	17155.418	25	17156.912	45	17158.709	65	17160.354
6	17155.523	26	17157.003	46	17158.815	66	17160.435
7	17155.591	27	17157.072	47	17158.872	67	17160.499
8	17155.723	28	17157.164	48	17158.959	68	17160.600
9	17155.826	29	17157.236	49	17159.049	69	17160.669
10	17155.899	30	17157.351	50	17159.109	70	17160.738
11	17155.942	31	17157.505	51	17159.188	71	17160.792
12	17156.045	32	17157.589	52	17159.380	72	17160.849
13	17156.108	33	17157.652	53	17159.415	73	17160.906
14	17156.155	34	17157.738	54	17159.503	74	17160.993
15	17156.229	35	17157.844	55	17159.563	75	17161.059
16	17156.309	36	17157.909	56	17159.609	76	17161.124
17	17156.399	37	17157.962	57	17159.713	77	17161.207
18	17156.505	38	17158.020	58	17159.822	78	17161.293
19	17156.561	39	17158.124	59	17159.933	79	17161.402
20	17156.615	40	17158.220	60	17159.987		

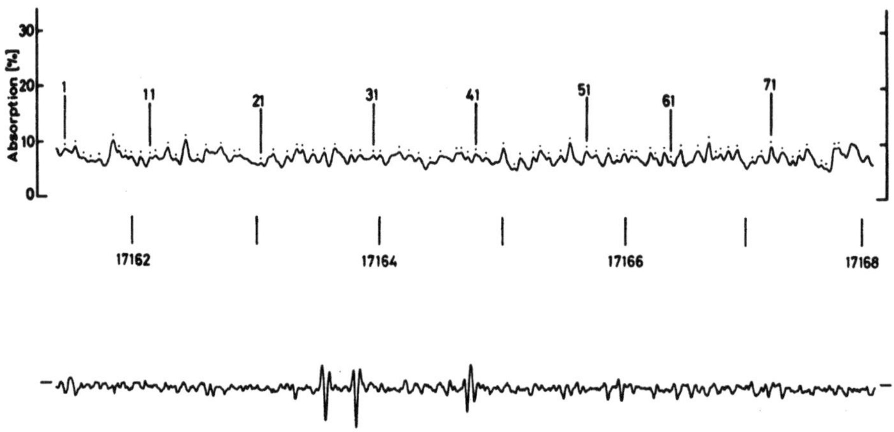

1	17161.460	21	17163.035	41	17164.778	61	17166.386
2	17161.541	22	17163.129	42	17164.867	62	17166.468
3	17161.616	23	17163.240	43	17165.008	63	17166.607
4	17161.672	24	17163.318	44	17165.098	64	17166.700
5	17161.727	25	17163.360	45	17165.147	65	17166.755
6	17161.846	26	17163.445	46	17165.251	66	17166.798
7	17161.899	27	17163.538	47	17165.310	67	17166.861
8	17161.954	28	17163.628	48	17165.382	68	17166.933
9	17161.998	29	17163.761	49	17165.476	69	17167.059
10	17162.074	30	17163.831	50	17165.550	70	17167.137
11	17162.150	31	17163.941	51	17165.681	71	17167.222
12	17162.195	32	17164.001	52	17165.760	72	17167.319
13	17162.288	33	17164.154	53	17165.863	73	17167.403
14	17162.354	34	17164.236	54	17165.932	74	17167.458
15	17162.429	35	17164.322	55	17165.996	75	17167.526
16	17162.530	36	17164.410	56	17166.058	76	17167.647
17	17162.592	37	17164.497	57	17166.107	77	17167.698
18	17162.711	38	17164.623	58	17166.213	78	17167.758
19	17162.818	39	17164.660	59	17166.264	79	17167.793
20	17162.863	40	17164.718	60	17166.334		

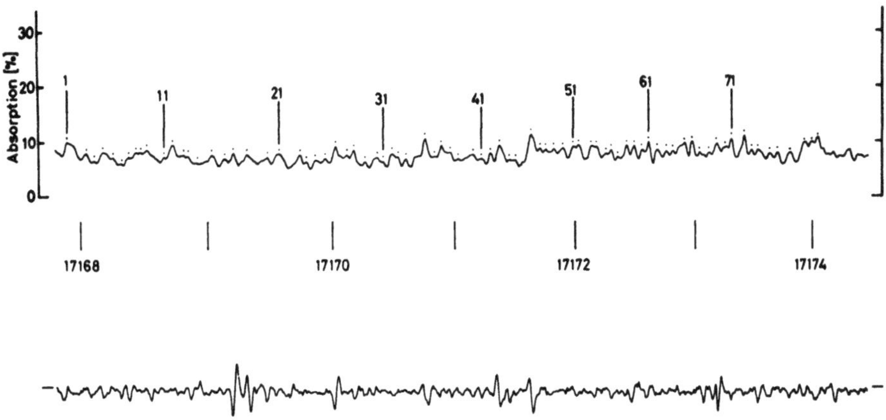

1	17167.900	21	17169.556	41	17171.222	61	17172.612
2	17168.046	22	17169.734	42	17171.294	62	17172.686
3	17168.107	23	17169.787	43	17171.367	63	17172.762
4	17168.180	24	17169.859	44	17171.449	64	17172.814
5	17168.254	25	17169.938	45	17171.504	65	17172.913
6	17168.316	26	17170.023	46	17171.630	66	17172.975
7	17168.368	27	17170.117	47	17171.705	67	17173.039
8	17168.430	28	17170.173	48	17171.755	68	17173.121
9	17168.465	29	17170.264	49	17171.821	69	17173.190
10	17168.510	30	17170.368	50	17171.898	70	17173.263
11	17168.649	31	17170.414	51	17171.981	71	17173.314
12	17168.718	32	17170.486	52	17172.029	72	17173.431
13	17168.809	33	17170.534	53	17172.135	73	17173.489
14	17168.849	34	17170.598	54	17172.177	74	17173.547
15	17168.900	35	17170.757	55	17172.256	75	17173.642
16	17169.031	36	17170.837	56	17172.299	76	17173.705
17	17169.127	37	17170.891	57	17172.353	77	17173.814
18	17169.195	38	17170.966	58	17172.429	78	17173.940
19	17169.304	39	17171.028	59	17172.492	79	17173.999
20	17169.462	40	17171.149	60	17172.558	80	17174.054

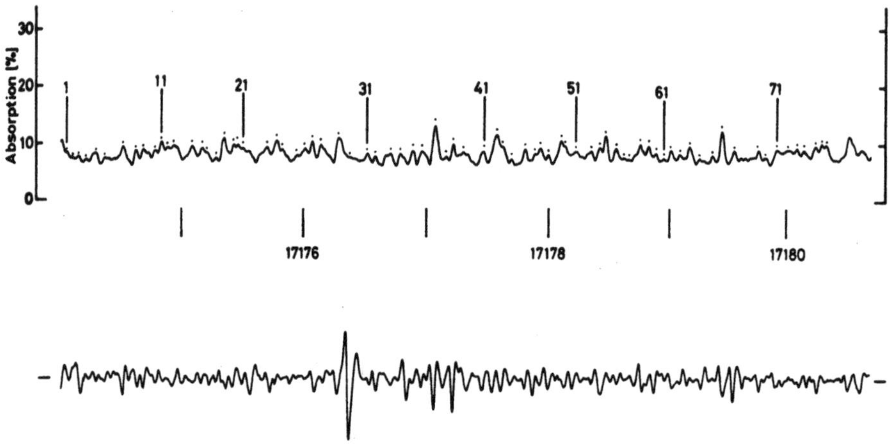

1	17174.099	21	17175.526	41	17177.465	61	17178.954
2	17174.140	22	17175.715	42	17177.574	62	17179.017
3	17174.188	23	17175.789	43	17177.633	63	17179.093
4	17174.248	24	17175.837	44	17177.693	64	17179.177
5	17174.323	25	17175.951	45	17177.808	65	17179.256
6	17174.396	26	17176.015	46	17177.891	66	17179.369
7	17174.549	27	17176.075	47	17177.936	67	17179.451
8	17174.655	28	17176.148	48	17177.998	68	17179.553
9	17174.715	29	17176.298	49	17178.102	69	17179.764
10	17174.809	30	17176.422	50	17178.140	70	17179.827
11	17174.859	31	17176.528	51	17178.221	71	17179.931
12	17174.906	32	17176.588	52	17178.349	72	17180.023
13	17174.959	33	17176.712	53	17178.422	73	17180.095
14	17175.099	34	17176.791	54	17178.472	74	17180.153
15	17175.180	35	17176.888	55	17178.558	75	17180.251
16	17175.222	36	17176.963	56	17178.618	76	17180.301
17	17175.295	37	17177.073	57	17178.665	77	17180.345
18	17175.362	38	17177.156	58	17178.759		
19	17175.446	39	17177.217	59	17178.829		
20	17175.481	40	17177.305	60	17178.886		

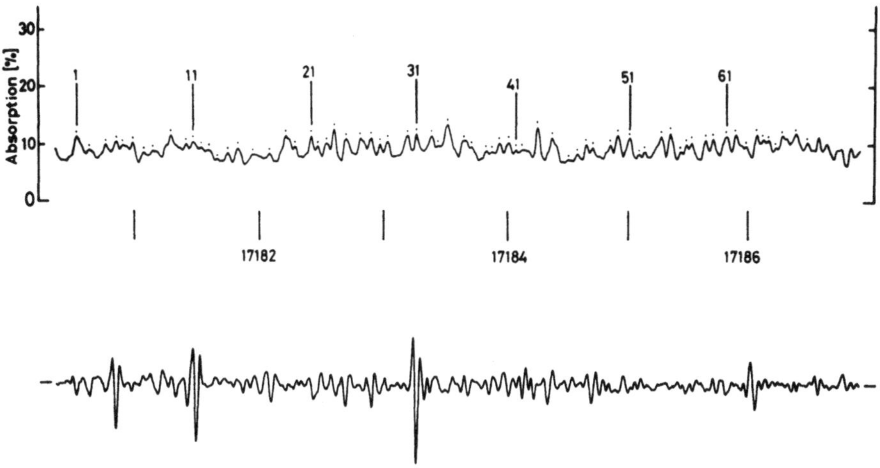

1	17180.535	21	17182.420	41	17184.072	61	17185.823
2	17180.641	22	17182.472	42	17184.126	62	17185.902
3	17180.769	23	17182.544	43	17184.251	63	17185.995
4	17180.858	24	17182.605	44	17184.373	64	17186.077
5	17180.910	25	17182.703	45	17184.504	65	17186.128
6	17180.991	26	17182.818	46	17184.579	66	17186.169
7	17181.082	27	17182.904	47	17184.654	67	17186.236
8	17181.158	28	17182.975	48	17184.706	68	17186.287
9	17181.301	29	17183.035	49	17184.852	69	17186.410
10	17181.419	30	17183.192	50	17184.911	70	17186.510
11	17181.473	31	17183.264	51	17185.010		
12	17181.537	32	17183.387	52	17185.083		
13	17181.596	33	17183.440	53	17185.136		
14	17181.661	34	17183.513	54	17185.272		
15	17181.748	35	17183.647	55	17185.348		
16	17181.829	36	17183.712	56	17185.436		
17	17181.946	37	17183.832	57	17185.487		
18	17182.082	38	17183.879	58	17185.538		
19	17182.215	39	17183.936	59	17185.653		
20	17182.287	40	17184.016	60	17185.717		

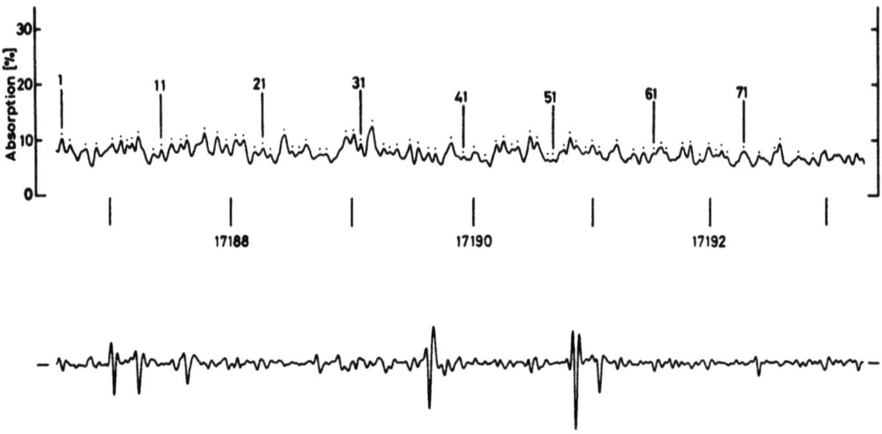

1	17186.610	21	17188.267	41	17189.917	61	17191.516
2	17186.675	22	17188.326	42	17190.002	62	17191.580
3	17186.800	23	17188.443	43	17190.100	63	17191.624
4	17186.889	24	17188.510	44	17190.188	64	17191.767
5	17187.021	25	17188.563	45	17190.246	65	17191.835
6	17187.098	26	17188.626	46	17190.321	66	17191.914
7	17187.151	27	17188.748	47	17190.365	67	17191.992
8	17187.190	28	17188.787	48	17190.468	68	17192.056
9	17187.243	29	17188.955	49	17190.526	69	17192.100
10	17187.369	30	17189.019	50	17190.612	70	17192.153
11	17187.431	31	17189.075	51	17190.662	71	17192.286
12	17187.512	32	17189.168	52	17190.752	72	17192.423
13	17187.590	33	17189.263	53	17190.807	73	17192.559
14	17187.643	34	17189.310	54	17190.861	74	17192.601
15	17187.788	35	17189.367	55	17190.945	75	17192.762
16	17187.895	36	17189.479	56	17190.997	76	17192.894
17	17187.969	37	17189.549	57	17191.057		
18	17188.049	38	17189.634	58	17191.198		
19	17188.111	39	17189.693	59	17191.342		
20	17188.200	40	17189.822	60	17191.429		

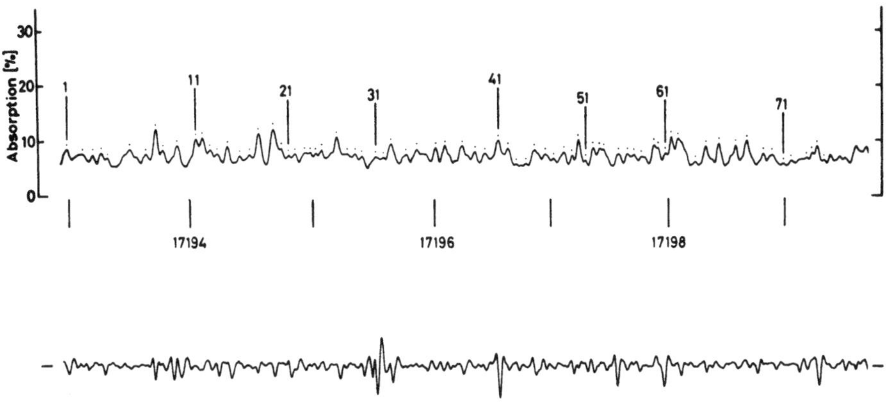

1	17192.990	21	17194.818	41	17196.551	61	17197.972
2	17193.130	22	17194.870	42	17196.633	62	17198.022
3	17193.209	23	17194.948	43	17196.703	63	17198.083
4	17193.280	24	17194.995	44	17196.776	64	17198.236
5	17193.330	25	17195.040	45	17196.856	65	17198.324
6	17193.510	26	17195.087	46	17196.952	66	17198.440
7	17193.644	27	17195.210	47	17197.023	67	17198.582
8	17193.731	28	17195.295	48	17197.118	68	17198.675
9	17193.786	29	17195.357	49	17197.183	69	17198.811
10	17193.907	30	17195.413	50	17197.240	70	17198.886
11	17194.057	31	17195.526	51	17197.298	71	17198.983
12	17194.112	32	17195.593	52	17197.365	72	17199.053
13	17194.174	33	17195.655	53	17197.422	73	17199.196
14	17194.229	34	17195.779	54	17197.457	74	17199.243
15	17194.315	35	17195.866	55	17197.572	75	17199.284
16	17194.414	36	17196.022	56	17197.651	76	17199.345
17	17194.495	37	17196.099	57	17197.705		
18	17194.574	38	17196.246	58	17197.769		
19	17194.693	39	17196.354	59	17197.872		
20	17194.760	40	17196.439	60	17197.906		

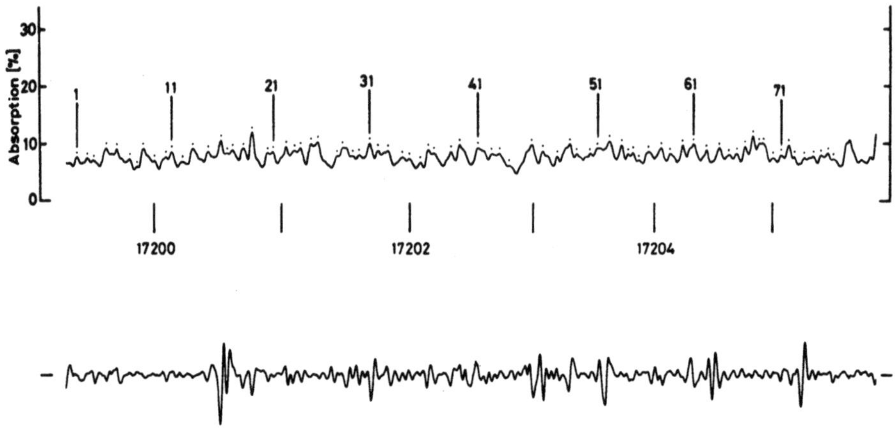

1	17199.398	21	17200.943	41	17202.555	61	17204.328
2	17199.476	22	17201.042	42	17202.672	62	17204.435
3	17199.526	23	17201.107	43	17202.735	63	17204.544
4	17199.624	24	17201.151	44	17202.824	64	17204.621
5	17199.712	25	17201.235	45	17202.992	65	17204.686
6	17199.814	26	17201.289	46	17203.085	66	17204.761
7	17199.870	27	17201.439	47	17203.201	67	17204.828
8	17199.926	28	17201.487	48	17203.301	68	17204.886
9	17200.011	29	17201.562	49	17203.366	69	17204.924
10	17200.099	30	17201.618	50	17203.475	70	17205.005
11	17200.147	31	17201.698	51	17203.531	71	17205.070
12	17200.235	32	17201.756	52	17203.632	72	17205.137
13	17200.318	33	17201.832	53	17203.729	73	17205.184
14	17200.438	34	17201.938	54	17203.787	74	17205.263
15	17200.538	35	17202.000	55	17203.829	75	17205.339
16	17200.587	36	17202.086	56	17203.874	76	17205.405
17	17200.633	37	17202.153	57	17203.957	77	17205.460
18	17200.716	38	17202.204	58	17204.061	78	17205.513
19	17200.786	39	17202.336	59	17204.138		
20	17200.903	40	17202.409	60	17204.239		

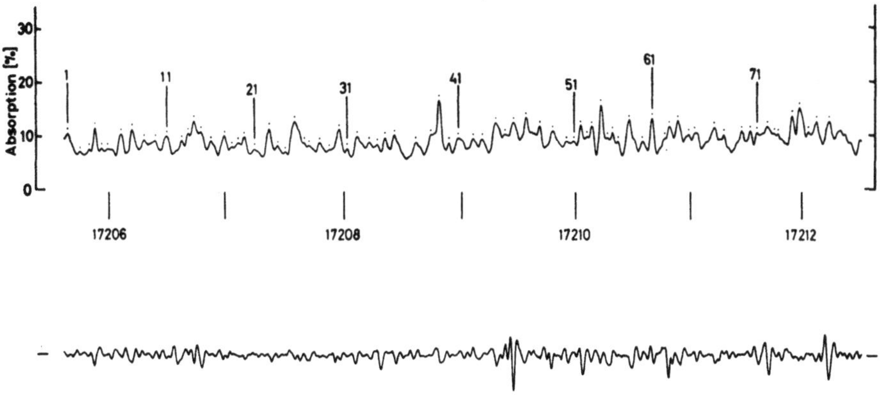

1	17205.642	21	17207.256	41	17208.978	61	17210.667
2	17205.746	22	17207.379	42	17209.099	62	17210.777
3	17205.822	23	17207.453	43	17209.177	63	17210.817
4	17205.878	24	17207.520	44	17209.295	64	17210.896
5	17205.939	25	17207.599	45	17209.384	65	17210.988
6	17205.986	26	17207.676	46	17209.448	66	17211.051
7	17206.105	27	17207.715	47	17209.559	67	17211.219
8	17206.206	28	17207.799	48	17209.682	68	17211.299
9	17206.305	29	17207.874	49	17209.802	69	17211.459
10	17206.410	30	17207.965	50	17209.919	70	17211.536
11	17206.506	31	17208.030	51	17209.983	71	17211.600
12	17206.641	32	17208.114	52	17210.050	72	17211.686
13	17206.692	33	17208.210	53	17210.104	73	17211.751
14	17206.747	34	17208.278	54	17210.147	74	17211.789
15	17206.800	35	17208.345	55	17210.227	75	17211.913
16	17206.885	36	17208.426	56	17210.285	76	17211.983
17	17206.998	37	17208.606	57	17210.323	77	17212.066
18	17207.065	38	17208.733	58	17210.374	78	17212.140
19	17207.111	39	17208.809	59	17210.471	79	17212.257
20	17207.169	40	17208.890	60	17210.584		

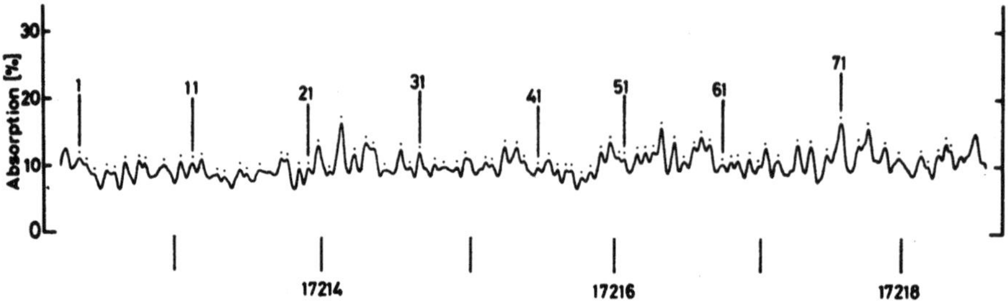

1	17212.354	21	17213.912	41	17215.460	61	17216.737
2	17212.403	22	17213.979	42	17215.532	62	17216.799
3	17212.460	23	17214.056	43	17215.598	63	17216.845
4	17212.539	24	17214.131	44	17215.657	64	17216.921
5	17212.590	25	17214.219	45	17215.690	65	17216.986
6	17212.668	26	17214.303	46	17215.765	66	17217.043
7	17212.769	27	17214.353	47	17215.828	67	17217.120
8	17212.811	28	17214.426	48	17215.904	68	17217.264
9	17212.933	29	17214.536	49	17215.964	69	17217.356
10	17213.045	30	17214.595	50	17216.030	70	17217.469
11	17213.123	31	17214.659	51	17216.064	71	17217.566
12	17213.184	32	17214.757	52	17216.152	72	17217.694
13	17213.285	33	17214.818	53	17216.211	73	17217.766
14	17213.338	34	17214.903	54	17216.266	74	17217.885
15	17213.445	35	17214.971	55	17216.320	75	17217.985
16	17213.495	36	17215.102	56	17216.409	76	17218.143
17	17213.589	37	17215.144	57	17216.471	77	17218.189
18	17213.735	38	17215.232	58	17216.546	78	17218.269
19	17213.769	39	17215.315	59	17216.590	79	17218.325
20	17213.851	40	17215.368	60	17216.653	80	17218.359

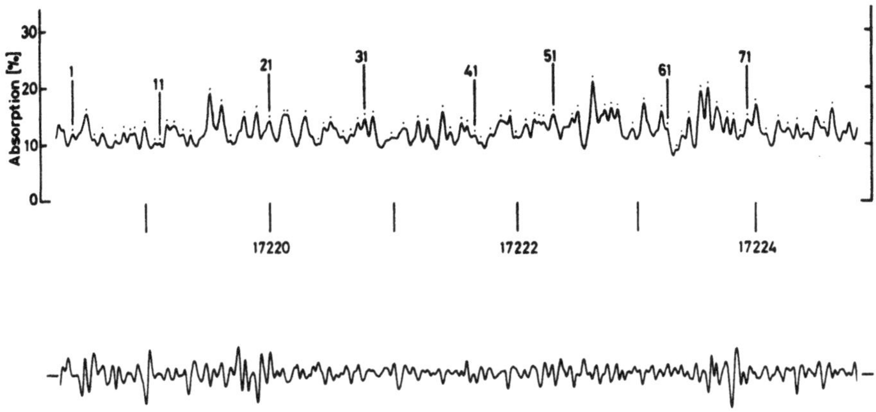

1	17218.430	21	17220.001	41	17221.667	61	17223.251
2	17218.538	22	17220.115	42	17221.717	62	17223.328
3	17218.604	23	17220.144	43	17221.794	63	17223.379
4	17218.666	24	17220.291	44	17221.876	64	17223.438
5	17218.767	25	17220.399	45	17221.950	65	17223.537
6	17218.830	26	17220.447	46	17221.999	66	17223.602
7	17218.878	27	17220.498	47	17222.082	67	17223.678
8	17218.914	28	17220.598	48	17222.152	68	17223.762
9	17218.993	29	17220.661	49	17222.188	69	17223.814
10	17219.067	30	17220.719	50	17222.226	70	17223.809
11	17219.108	31	17220.772	51	17222.307	71	17223.936
12	17219.176	32	17220.839	52	17222.460	72	17224.006
13	17219.219	33	17220.992	53	17222.506	73	17224.084
14	17219.298	34	17221.089	54	17222.631	74	17224.195
15	17219.362	35	17221.208	55	17222.729	75	17224.278
16	17219.514	36	17221.285	56	17222.783	76	17224.357
17	17219.606	37	17221.403	57	17222.837	77	17224.419
18	17219.681	38	17221.467	58	17222.960	78	17224.518
19	17219.791	39	17221.557	59	17223.055	79	17224.584
20	17219.898	40	17221.604	60	17223.203	80	17224.654

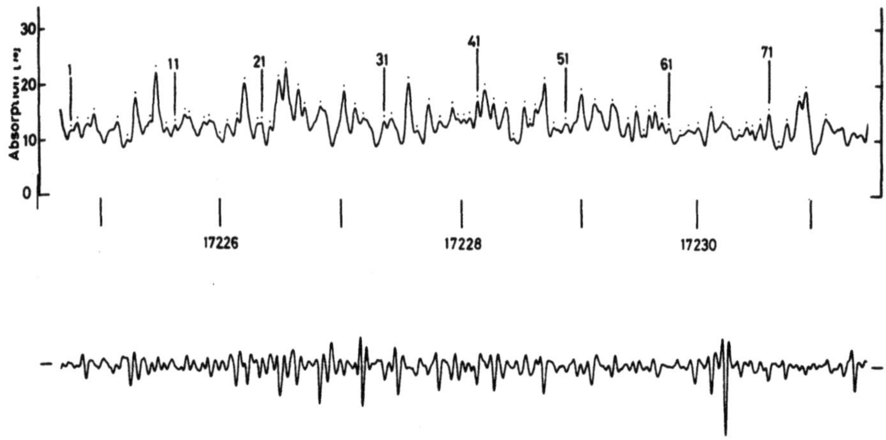

1	17224.753	21	17226.341	41	17228.136	61	17229.756
2	17224.804	22	17226.413	42	17228.196	62	17229.852
3	17224.894	23	17226.486	43	17228.269	63	17229.922
4	17224.950	24	17226.542	44	17228.368	64	17230.006
5	17225.147	25	17226.643	45	17228.435	65	17230.119
6	17225.236	26	17226.699	46	17228.521	66	17230.221
7	17225.307	27	17226.835	47	17228.574	67	17230.366
8	17225.424	28	17227.025	48	17228.619	68	17230.430
9	17225.479	29	17227.123	49	17228.689	69	17230.476
10	17225.567	30	17227.187	50	17228.775	70	17230.552
11	17225.635	31	17227.365	51	17228.864	71	17230.625
12	17225.716	32	17227.422	52	17229.000	72	17230.714
13	17225.755	33	17227.563	53	17229.116	73	17230.787
14	17225.873	34	17227.640	54	17229.269	74	17230.890
15	17225.921	35	17227.730	55	17229.403	75	17230.948
16	17226.004	36	17227.821	56	17229.474	76	17231.122
17	17226.069	37	17227.928	57	17229.538		
18	17226.153	38	17227.975	58	17229.584		
19	17226.203	39	17228.019	59	17229.637		
20	17226.313	40	17228.074	60	17229.697		

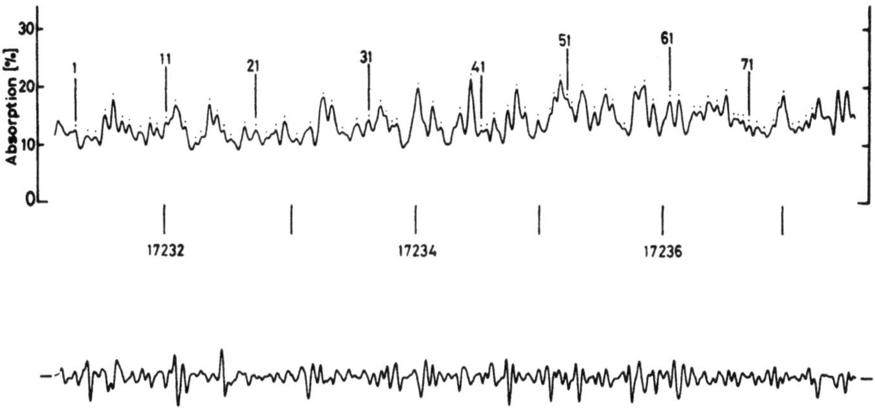

1	17231.268	21	17232.722	41	17234.531	61	17236.059
2	17231.370	22	17232.807	42	17234.578	62	17236.140
3	17231.440	23	17232.881	43	17234.635	63	17236.261
4	17231.516	24	17232.951	44	17234.743	64	17236.319
5	17231.588	25	17233.045	45	17234.816	65	17236.395
6	17231.656	26	17233.155	46	17234.886	66	17236.456
7	17231.713	27	17233.258	47	17234.991	67	17236.528
8	17231.804	28	17233.328	48	17235.089	68	17236.582
9	17231.888	29	17233.411	49	17235.125	69	17236.624
10	17231.942	30	17233.527	50	17235.172	70	17236.668
11	17232.012	31	17233.623	51	17235.227	71	17236.719
12	17232.092	32	17233.718	52	17235.265	72	17236.789
13	17232.162	33	17233.761	53	17235.349	73	17236.840
14	17232.248	34	17233.813	54	17235.452	74	17236.918
15	17232.304	35	17233.849	55	17235.542	75	17236.969
16	17232.358	36	17234.023	56	17235.595	76	17237.007
17	17232.420	37	17234.140	57	17235.693	77	17237.092
18	17232.468	38	17234.207	58	17235.781	78	17237.145
19	17232.521	39	17234.363	59	17235.854	79	17237.204
20	17232.634	40	17234.448	60	17235.925	80	17237.257

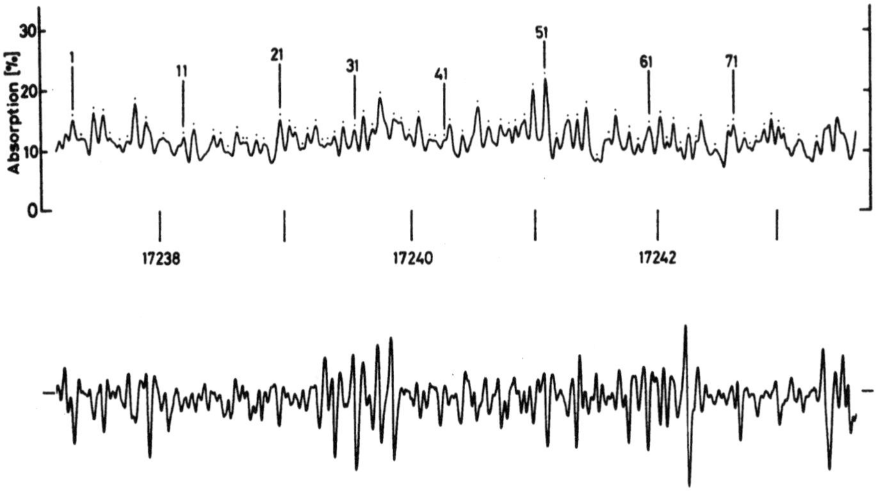

1	17237.313	21	17238.961	41	17240.260	61	17241.934
2	17237.388	22	17239.037	42	17240.303	62	17242.026
3	17237.475	23	17239.088	43	17240.413	63	17242.082
4	17237.556	24	17239.138	44	17240.534	64	17242.133
5	17237.605	25	17239.184	45	17240.619	65	17242.194
6	17237.684	26	17239.251	46	17240.717	66	17242.259
7	17237.746	27	17239.298	47	17240.780	67	17242.324
8	17237.806	28	17239.342	48	17240.839	68	17242.366
9	17237.894	29	17239.392	49	17240.909	69	17242.489
10	17238.027	30	17239.464	50	17240.985	70	17242.591
11	17238.190	31	17239.551	51	17241.086	71	17242.636
12	17238.270	32	17239.620	52	17241.174	72	17242.729
13	17238.431	33	17239.690	53	17241.270	73	17242.770
14	17238.492	34	17239.754	54	17241.343	74	17242.823
15	17238.557	35	17239.865	55	17241.422	75	17242.887
16	17238.621	36	17239.911	56	17241.514	76	17242.954
17	17238.661	37	17239.979	57	17241.603	77	17243.014
18	17238.698	38	17240.056	58	17241.661	78	17243.052
19	17238.775	39	17240.147	59	17241.773	79	17243.191
20	17238.845	40	17240.188	60	17241.846	80	17243.334

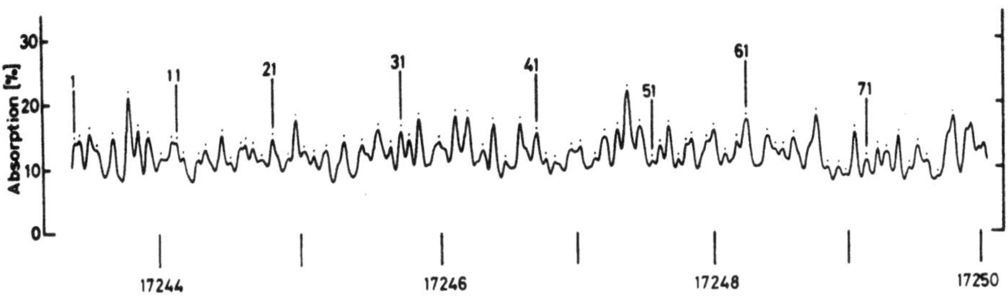

1	17243.401	21	17244.805	41	17246.707	61	17248.248
2	17243.436	22	17244.917	42	17246.775	62	17248.402
3	17243.505	23	17244.972	43	17246.839	63	17248.462
4	17243.566	24	17245.031	44	17246.960	64	17248.513
5	17243.678	25	17245.100	45	17247.023	65	17248.596
6	17243.795	26	17245.187	46	17247.130	66	17248.768
7	17243.860	27	17245.313	47	17247.202	67	17248.848
8	17243.929	28	17245.442	48	17247.295	68	17248.931
9	17244.025	29	17245.558	49	17247.368	69	17248.985
10	17244.099	30	17245.649	50	17247.458	70	17249.048
11	17244.134	31	17245.721	51	17247.551	71	17249.133
12	17244.178	32	17245.780	52	17247.608	72	17249.216
13	17244.285	33	17245.852	53	17247.672	73	17249.292
14	17244.334	34	17245.999	54	17247.745	74	17249.374
15	17244.451	35	17246.115	55	17247.801	75	17249.458
16	17244.513	36	17246.205	56	17247.839	76	17249.521
17	17244.582	37	17246.318	57	17247.956	77	17249.592
18	17244.617	38	17246.395	58	17248.002	78	17249.672
19	17244.668	39	17246.480	59	17248.086	79	17249.794
20	17244.731	40	17246.586	60	17248.173		

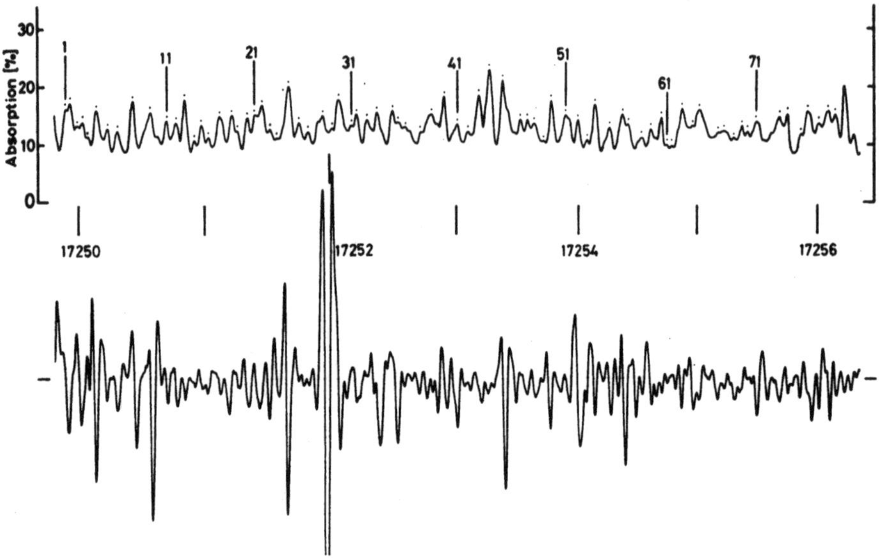

1	17249.893	21	17251.411	41	17253.013	61	17254.748
2	17249.929	22	17251.471	42	17253.106	62	17254.792
3	17249.991	23	17251.525	43	17253.198	63	17254.874
4	17250.032	24	17251.583	44	17253.277	64	17254.963
5	17250.074	25	17251.678	45	17253.384	65	17255.020
6	17250.142	26	17251.754	46	17253.520	66	17255.170
7	17250.234	27	17251.831	47	17253.581	67	17255.218
8	17250.318	28	17251.943	48	17253.634	68	17255.308
9	17250.443	29	17252.017	49	17253.712	69	17255.372
10	17250.584	30	17252.077	50	17253.777	70	17255.419
11	17250.710	31	17252.173	51	17253.898	71	17255.495
12	17250.786	32	17252.212	52	17253.997	72	17255.682
13	17250.851	33	17252.300	53	17254.074	73	17255.750
14	17250.927	34	17252.380	54	17254.144	74	17255.865
15	17250.982	35	17252.504	55	17254.262	75	17255.926
16	17251.042	36	17252.544	56	17254.372	76	17256.020
17	17251.135	37	17252.599	57	17254.418	77	17256.100
18	17251.234	38	17252.725	58	17254.533	78	17256.161
19	17251.285	39	17252.808	59	17254.609		
20	17251.352	40	17252.913	60	17254.700		

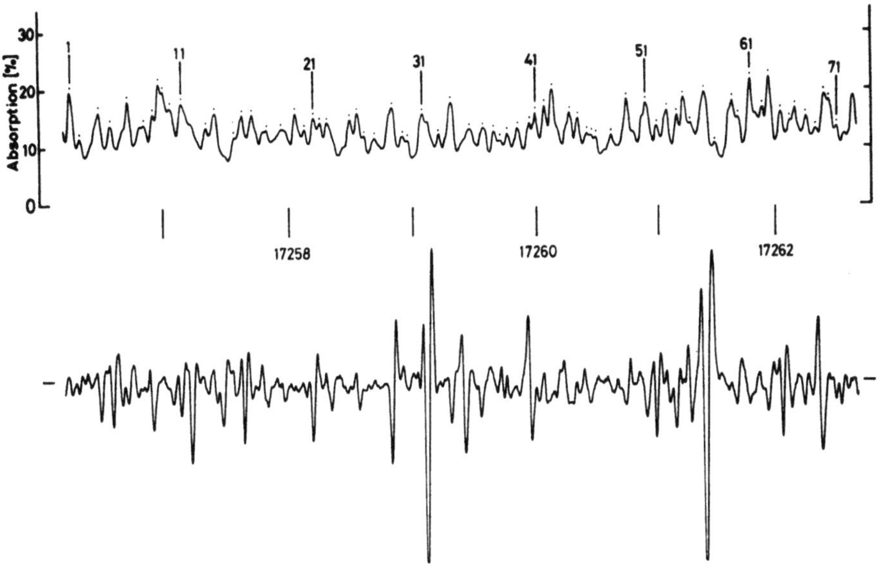

1	17256.240	21	17258.222	41	17259.988	61	17261.788
2	17256.319	22	17258.275	42	17260.066	62	17261.887
3	17256.473	23	17258.332	43	17260.131	63	17261.946
4	17256.568	24	17258.513	44	17260.272	64	17262.046
5	17256.715	25	17258.571	45	17260.335	65	17262.126
6	17256.842	26	17258.624	46	17260.406	66	17262.172
7	17256.923	27	17258.708	47	17260.450	67	17262.264
8	17256.971	28	17258.845	48	17260.484	68	17262.350
9	17257.006	29	17258.929	49	17260.616	69	17262.421
10	17257.062	30	17258.971	50	17260.746	70	17262.456
11	17257.154	31	17259.092	51	17260.908	71	17262.523
12	17257.352	32	17259.220	52	17260.989		
13	17257.423	33	17259.315	53	17261.072		
14	17257.578	34	17259.460	54	17261.169		
15	17257.642	35	17259.568	55	17261.220		
16	17257.723	36	17259.655	56	17261.272		
17	17257.845	37	17259.708	57	17261.408		
18	17257.963	38	17259.761	58	17261.490		
19	17258.076	39	17259.845	59	17261.634		
20	17258.152	40	17259.938	60	17261.688		

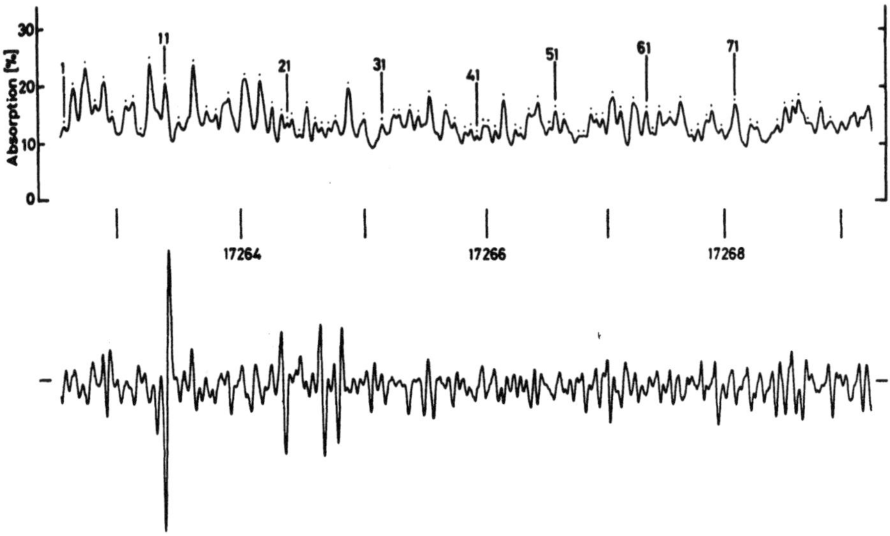

1	17262.590	21	17264.377	41	17265.916	61	17267.322
2	17262.664	22	17264.415	42	17265.971	62	17267.380
3	17262.757	23	17264.477	43	17266.009	63	17267.438
4	17262.831	24	17264.537	44	17266.070	64	17267.524
5	17262.900	25	17264.610	45	17266.139	65	17267.613
6	17262.957	26	17264.658	46	17266.237	66	17267.709
7	17263.076	27	17264.716	47	17266.285	67	17267.767
8	17263.133	28	17264.772	48	17266.348	68	17267.842
9	17263.193	29	17264.881	49	17266.422	69	17267.887
10	17263.273	30	17264.996	50	17266.519	70	17267.966
11	17263.396	31	17265.146	51	17266.564	71	17268.087
12	17263.504	32	17265.240	52	17266.637	72	17268.228
13	17263.621	33	17265.284	53	17266.759	73	17268.280
14	17263.723	34	17265.371	54	17266.799	74	17268.456
15	17263.797	35	17265.440	55	17266.860	75	17268.531
16	17263.902	36	17265.539	56	17266.905	76	17268.597
17	17264.040	37	17265.667	57	17266.956	77	17268.645
18	17264.162	38	17265.738	58	17267.038	78	17268.732
19	17264.256	39	17265.821	59	17267.106	79	17268.836
20	17264.331	40	17265.868	60	17267.219	80	17268.915

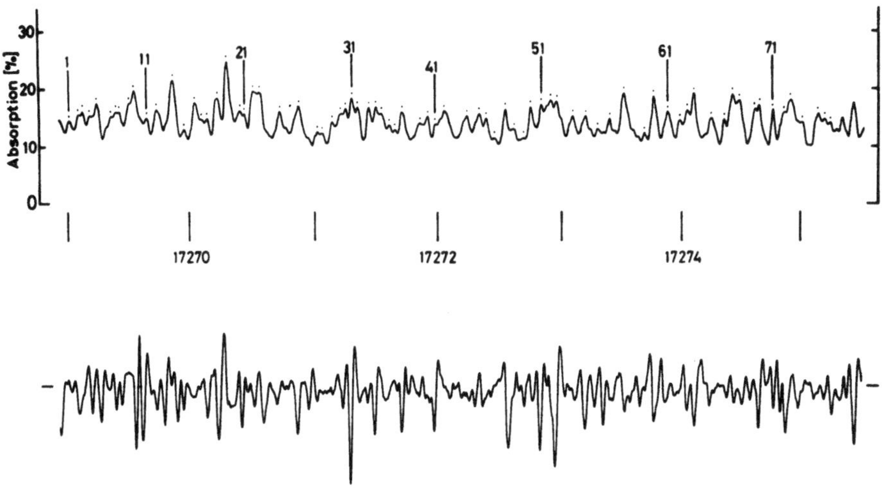

1	17269.011	21	17270.446	41	17271.981	61	17273.876
2	17269.080	22	17270.525	42	17272.063	62	17273.980
3	17269.117	23	17270.570	43	17272.234	63	17274.046
4	17269.180	24	17270.729	44	17272.335	64	17274.098
5	17269.238	25	17270.876	45	17272.388	65	17274.242
6	17269.355	26	17271.022	46	17272.460	66	17274.355
7	17269.393	27	17271.066	47	17272.546	67	17274.428
8	17269.425	28	17271.142	48	17272.619	68	17274.482
9	17269.506	29	17271.213	49	17272.687	69	17274.612
10	17269.554	30	17271.250	50	17272.758	70	17274.650
11	17269.651	31	17271.303	51	17272.843	71	17274.767
12	17269.735	32	17271.348	52	17272.920	72	17274.859
13	17269.872	33	17271.436	53	17272.962	73	17274.930
14	17269.962	34	17271.497	54	17273.096	74	17275.026
15	17270.048	35	17271.545	55	17273.203	75	17275.166
16	17270.106	36	17271.602	56	17273.305	76	17275.221
17	17270.149	37	17271.651	57	17273.399	77	17275.268
18	17270.224	38	17271.708	58	17273.518		
19	17270.304	39	17271.855	59	17273.687		
20	17270.409	40	17271.918	60	17273.762		

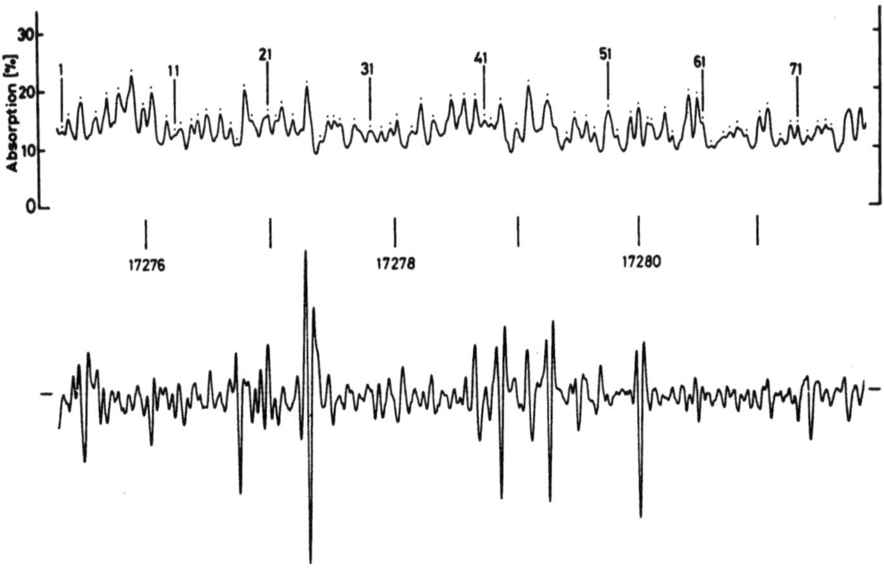

1	17275.324	21	17276.992	41	17278.732	61	17280.544
2	17275.375	22	17277.062	42	17278.793	62	17280.626
3	17275.478	23	17277.106	43	17278.865	63	17280.721
4	17275.598	24	17277.196	44	17278.991	64	17280.772
5	17275.691	25	17277.317	45	17279.094	65	17280.840
6	17275.793	26	17277.417	46	17279.249	66	17280.915
7	17275.896	27	17277.484	47	17279.404	67	17281.027
8	17275.988	28	17277.531	48	17279.473	68	17281.105
9	17276.055	29	17277.577	49	17279.574	69	17281.209
10	17276.178	30	17277.689	50	17279.643	70	17281.288
11	17276.239	31	17277.817	51	17279.762	71	17281.353
12	17276.287	32	17277.901	52	17279.828	72	17281.435
13	17276.376	33	17277.974	53	17279.939	73	17281.523
14	17276.430	34	17278.025	54	17280.009	74	17281.584
15	17276.501	35	17278.141	55	17280.079	75	17281.630
16	17276.619	36	17278.215	56	17280.121		
17	17276.704	37	17278.313	57	17280.228		
18	17276.753	38	17278.463	58	17280.289		
19	17276.814	39	17278.565	59	17280.431		
20	17276.872	40	17278.659	60	17280.501		

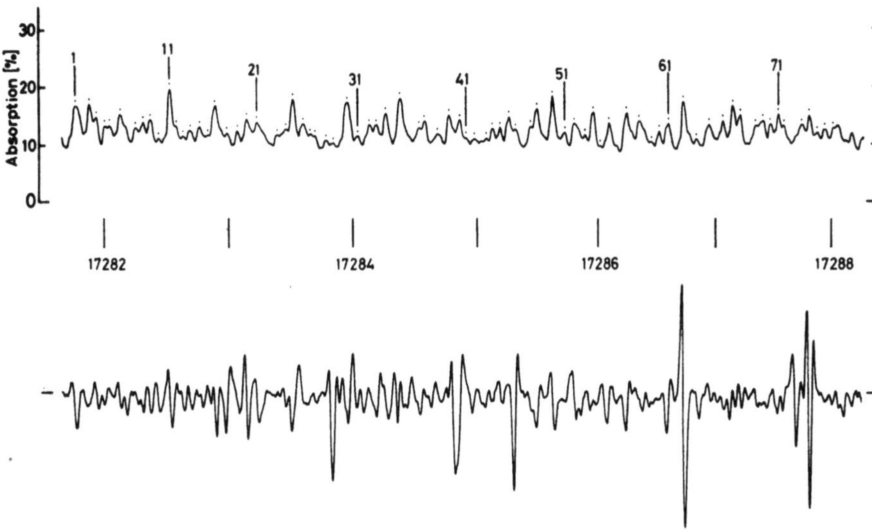

1	17281.790	21	17283.231	41	17284.921	61	17286.603
2	17281.886	22	17283.396	42	17284.981	62	17286.738
3	17281.932	23	17283.459	43	17285.076	63	17286.845
4	17282.009	24	17283.522	44	17285.130	64	17286.954
5	17282.042	25	17283.605	45	17285.198	65	17287.072
6	17282.122	26	17283.665	46	17285.268	66	17287.160
7	17282.250	27	17283.708	47	17285.315	67	17287.222
8	17282.304	28	17283.792	48	17285.442	68	17287.346
9	17282.363	29	17283.845	49	17285.492	69	17287.416
10	17282.431	30	17283.961	50	17285.621	70	17287.478
11	17282.525	31	17284.041	51	17285.720	71	17287.544
12	17282.570	32	17284.135	52	17285.805	72	17287.589
13	17282.623	33	17284.195	53	17285.887	73	17287.753
14	17282.689	34	17284.265	54	17285.966	74	17287.811
15	17282.764	35	17284.386	55	17286.033	75	17287.884
16	17282.817	36	17284.528	56	17286.097	76	17287.947
17	17282.890	37	17284.585	57	17286.248	77	17288.012
18	17282.985	38	17284.678	58	17286.356		
19	17283.069	39	17284.775	59	17286.449		
20	17283.143	40	17284.867	60	17286.527		

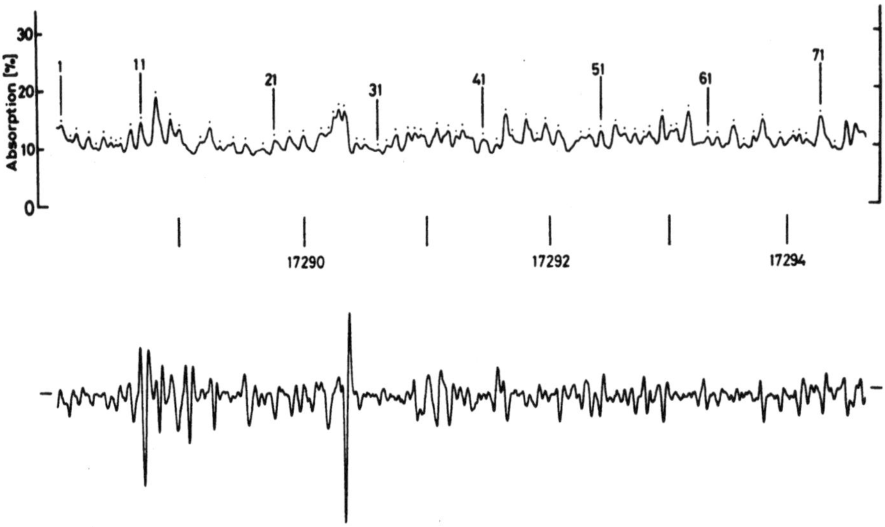

1	17288.049	21	17289.778	41	17291.474	61	17293.322
2	17288.127	22	17289.890	42	17291.573	62	17293.401
3	17288.173	23	17290.001	43	17291.646	63	17293.535
4	17288.267	24	17290.149	44	17291.705	64	17293.621
5	17288.329	25	17290.212	45	17291.810	65	17293.705
6	17288.397	26	17290.261	46	17291.898	66	17293.785
7	17288.458	27	17290.294	47	17291.968	67	17293.939
8	17288.502	28	17290.351	48	17292.077	68	17294.051
9	17288.536	29	17290.435	49	17292.259	69	17294.108
10	17288.619	30	17290.505	50	17292.329	70	17294.166
11	17288.706	31	17290.623	51	17292.422	71	17294.288
12	17288.828	32	17290.683	52	17292.547	72	17294.411
13	17288.946	33	17290.757	53	17292.621		
14	17289.013	34	17290.854	54	17292.711		
15	17289.183	35	17290.907	55	17292.789		
16	17289.260	36	17290.975	56	17292.826		
17	17289.341	37	17291.086	57	17292.937		
18	17289.445	38	17291.175	58	17293.015		
19	17289.543	39	17291.245	59	17293.058		
20	17289.678	40	17291.299	60	17293.159		

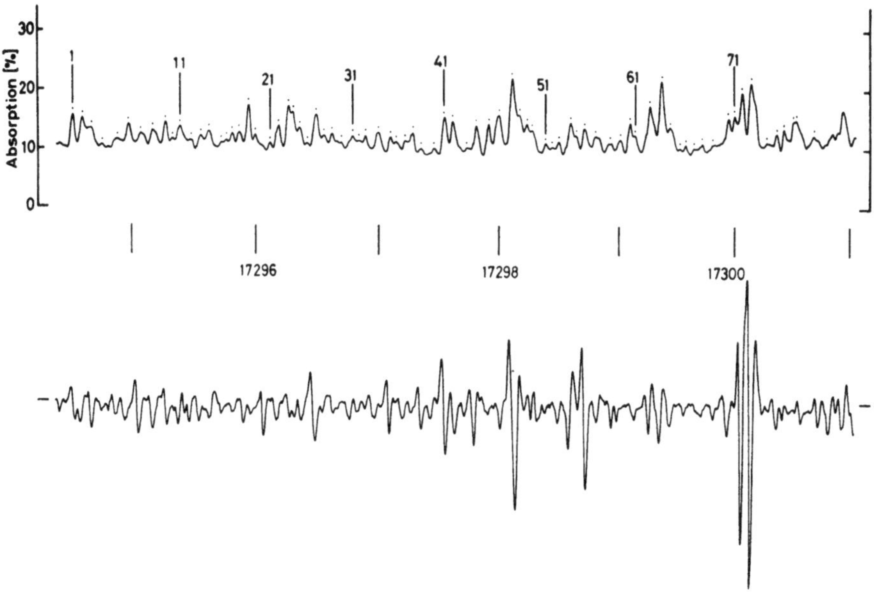

1	17294.519	21	17296.118	41	17297.546	61	17299.142
2	17294.598	22	17296.181	42	17297.619	62	17299.269
3	17294.667	23	17296.266	43	17297.737	63	17299.369
4	17294.757	24	17296.296	44	17297.812	64	17299.436
5	17294.891	25	17296.360	45	17297.917	65	17299.516
6	17294.977	26	17296.426	46	17298.001	66	17299.572
7	17295.084	27	17296.489	47	17298.118	67	17299.667
8	17295.177	28	17296.567	48	17298.173	68	17299.714
9	17295.279	29	17296.628	49	17298.237	69	17299.823
10	17295.329	30	17296.685	50	17298.292	70	17299.949
11	17295.394	31	17296.789	51	17298.394	71	17299.999
12	17295.476	32	17296.844	52	17298.446	72	17300.065
13	17295.555	33	17296.892	53	17298.505	73	17300.143
14	17295.624	34	17296.998	54	17298.601	74	17300.284
15	17295.724	35	17297.105	55	17298.646	75	17300.365
16	17295.766	36	17297.141	56	17298.717	76	17300.431
17	17295.812	37	17297.224	57	17298.808	77	17300.516
18	17295.865	38	17297.286	58	17298.934	78	17300.545
19	17295.942	39	17297.341	59	17299.010	79	17300.694
20	17295.996	40	17297.451	60	17299.096		

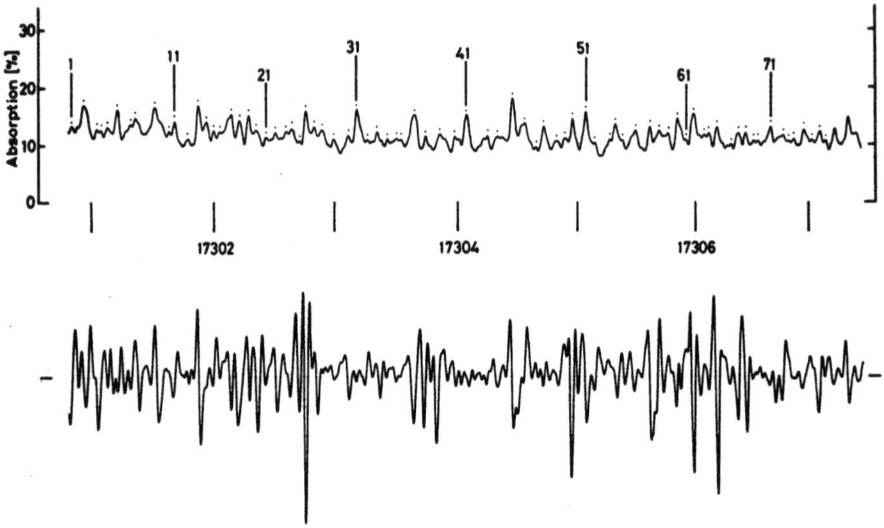

1	17300.852	21	17302.438	41	17304.077	61	17305.925
2	17300.949	22	17302.507	42	17304.246	62	17305.995
3	17301.048	23	17302.596	43	17304.334	63	17306.083
4	17301.076	24	17302.646	44	17304.464	64	17306.131
5	17301.137	25	17302.703	45	17304.561	65	17306.202
6	17301.222	26	17302.762	46	17304.661	66	17306.320
7	17301.327	27	17302.827	47	17304.720	67	17306.392
8	17301.370	28	17302.894	48	17304.825	68	17306.450
9	17301.531	29	17302.995	49	17304.901	69	17306.519
10	17301.639	30	17303.106	50	17304.962	70	17306.559
11	17301.688	31	17303.181	51	17305.076	71	17306.672
12	17301.789	32	17303.267	52	17305.152	72	17306.780
13	17301.883	33	17303.341	53	17305.266	73	17306.823
14	17301.943	34	17303.422	54	17305.319	74	17306.874
15	17302.004	35	17303.489	55	17305.384	75	17306.964
16	17302.058	36	17303.531	56	17305.492	76	17307.029
17	17302.148	37	17303.641	57	17305.616	77	17307.096
18	17302.222	38	17303.732	58	17305.696	78	17307.145
19	17302.296	39	17303.853	59	17305.775		
20	17302.346	40	17303.976	60	17305.860		

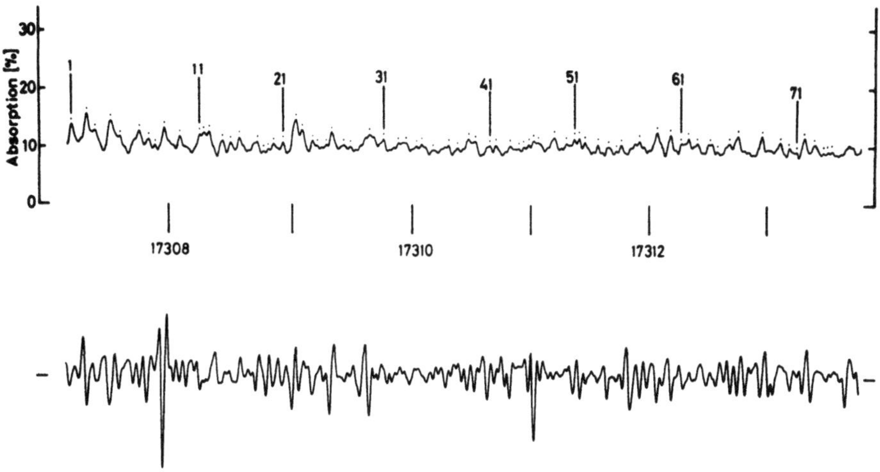

1	17307.220	21	17308.926	41	17310.659	61	17312.280
2	17307.338	22	17309.034	42	17310.710	62	17312.335
3	17307.402	23	17309.085	43	17310.830	63	17312.411
4	17307.534	24	17309.170	44	17310.903	64	17312.520
5	17307.604	25	17309.210	45	17310.945	65	17312.588
6	17307.763	26	17309.284	46	17310.985	66	17312.678
7	17307.836	27	17309.330	47	17311.028	67	17312.755
8	17307.887	28	17309.432	48	17311.068	68	17312.959
9	17307.967	29	17309.483	49	17311.193	69	17313.116
10	17308.091	30	17309.644	50	17311.303	70	17313.207
11	17308.249	31	17309.761	51	17311.355	71	17313.270
12	17308.285	32	17309.886	52	17311.398	72	17313.343
13	17308.324	33	17309.952	53	17311.456	73	17313.430
14	17308.437	34	17310.037	54	17311.562	74	17313.511
15	17308.502	35	17310.082	55	17311.664	75	17313.548
16	17308.573	36	17310.179	56	17311.714	76	17313.582
17	17308.716	37	17310.308	57	17311.763		
18	17308.767	38	17310.386	58	17311.915		
19	17308.808	39	17310.477	59	17312.068		
20	17308.850	40	17310.537	60	17312.180		

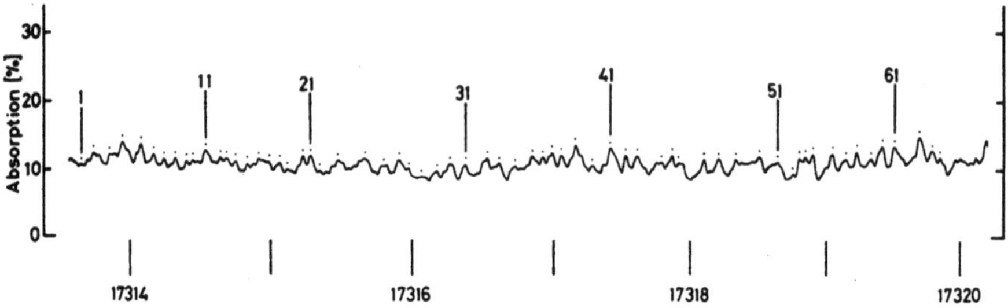

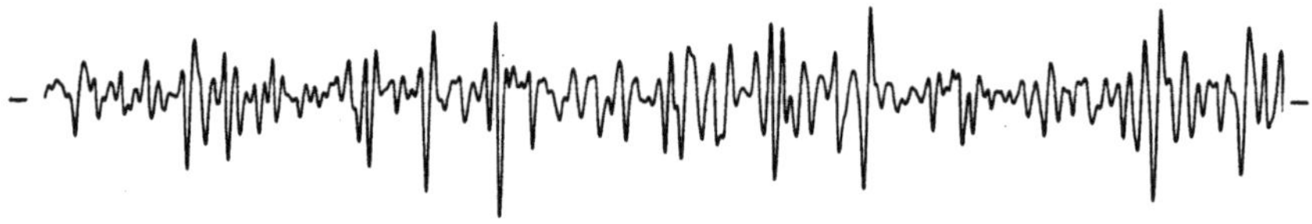

1	17313.635	21	17315.290	41	17317.418	61	17319.515
2	17313.723	22	17315.350	42	17317.527	62	17319.704
3	17313.860	23	17315.496	43	17317.612	63	17319.793
4	17313.946	24	17315.661	44	17317.792	64	17319.856
5	17314.083	25	17315.817	45	17317.863		
6	17314.177	26	17315.925	46	17317.933		
7	17314.242	27	17315.982	47	17318.105		
8	17314.319	28	17316.088	48	17318.211		
9	17314.408	29	17316.178	49	17318.343		
10	17314.458	30	17316.282	50	17318.514		
11	17314.543	31	17316.376	51	17318.638		
12	17314.645	32	17316.536	52	17318.751		
13	17314.695	33	17316.624	53	17318.806		
14	17314.757	34	17316.730	54	17318.847		
15	17314.843	35	17316.858	55	17318.900		
16	17314.932	36	17316.935	56	17319.041		
17	17314.990	37	17316.991	57	17319.138		
18	17315.065	38	17317.061	58	17319.223		
19	17315.145	39	17317.162	59	17319.330		
20	17315.231	40	17317.285	60	17319.412		

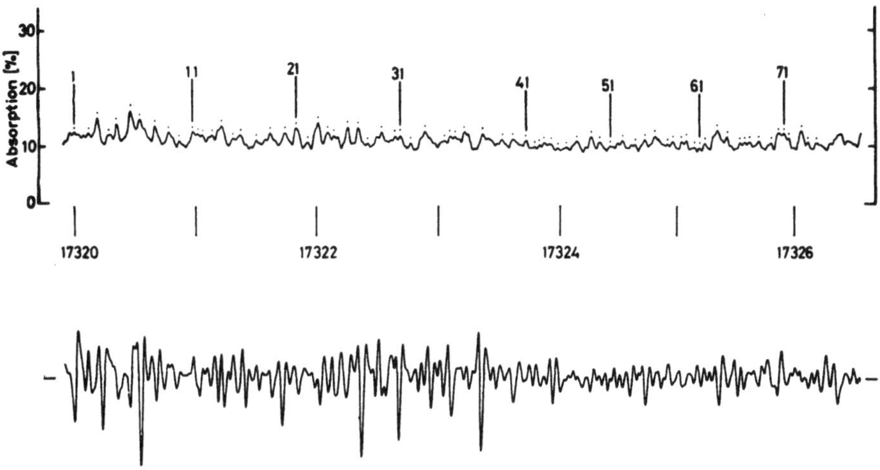

1	17319.997	21	17321.845	41	17323.717	61	17325.194
2	17320.131	22	17321.931	42	17323.792	62	17325.242
3	17320.209	23	17322.014	43	17323.834	63	17325.349
4	17320.309	24	17322.094	44	17323.871	64	17325.433
5	17320.367	25	17322.150	45	17323.929	65	17325.546
6	17320.480	26	17322.258	46	17323.987	66	17325.586
7	17320.556	27	17322.355	47	17324.055	67	17325.627
8	17320.681	28	17322.431	48	17324.144	68	17325.709
9	17320.788	29	17322.547	49	17324.265	69	17325.817
10	17320.879	30	17322.656	50	17324.337	70	17325.886
11	17320.990	31	17322.697	51	17324.427	71	17325.932
12	17321.025	32	17322.780	52	17324.529	72	17325.975
13	17321.062	33	17322.897	53	17324.639	73	17326.088
14	17321.142	34	17323.053	54	17324.721	74	17326.146
15	17321.222	35	17323.098	55	17324.807	75	17326.220
16	17321.322	36	17323.137	56	17324.916		
17	17321.379	37	17323.215	57	17324.962		
18	17321.519	38	17323.363	58	17325.038		
19	17321.630	39	17323.523	59	17325.081		
20	17321.748	40	17323.612	60	17325.149		

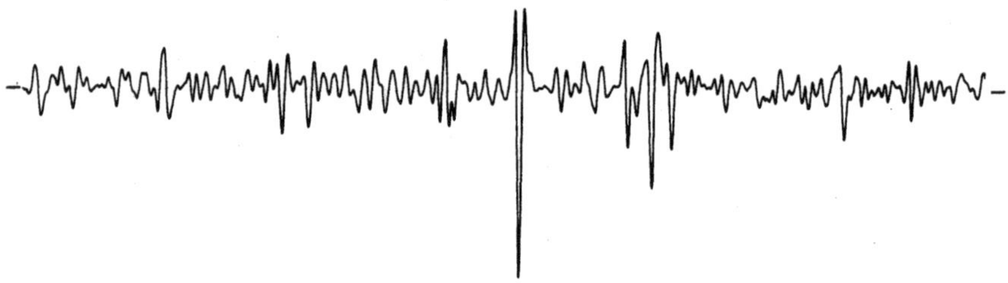

1	17326.425	21	17328.120	41	17330.306	61	17332.203
2	17326.482	22	17328.303	42	17330.385	62	17332.325
3	17326.607	23	17328.372	43	17330.459	63	17332.406
4	17326.729	24	17328.498	44	17330.535	64	17332.481
5	17326.795	25	17328.635	45	17330.617	65	17332.580
6	17326.837	26	17328.752	46	17330.695	66	17332.668
7	17326.943	27	17328.793	47	17330.812		
8	17326.999	28	17328.882	48	17330.858		
9	17327.050	29	17328.995	49	17331.021		
10	17327.132	30	17329.096	50	17331.060		
11	17327.225	31	17329.156	51	17331.152		
12	17327.302	32	17329.315	52	17331.277		
13	17327.473	33	17329.495	53	17331.458		
14	17327.521	34	17329.646	54	17331.588		
15	17327.576	35	17329.700	55	17331.631		
16	17327.683	36	17329.775	56	17331.762		
17	17327.737	37	17329.891	57	17331.899		
18	17327.806	38	17329.994	58	17331.968		
19	17327.943	39	17330.152	59	17332.085		
20	17328.062	40	17330.211	60	17332.148		

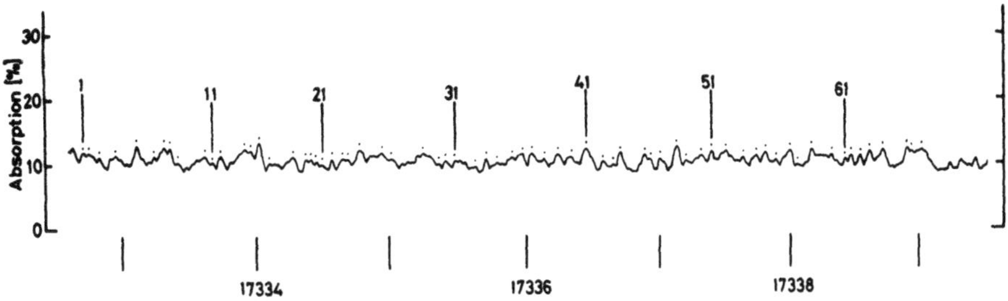

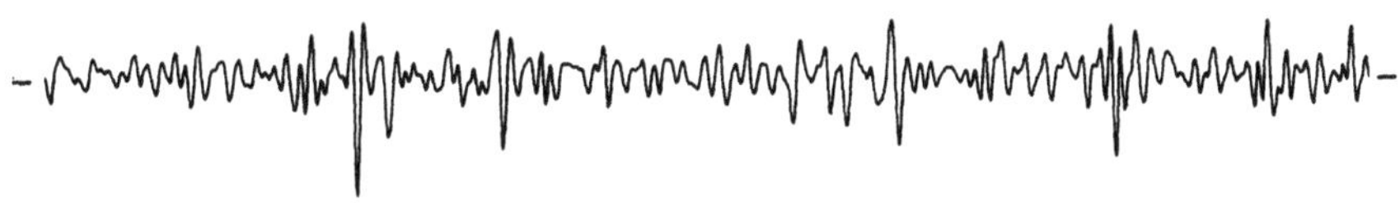

1	17332.751	21	17334.504	41	17336.445	61	17338.437
2	17332.797	22	17334.570	42	17336.576	62	17338.485
3	17332.861	23	17334.646	43	17336.645	63	17338.552
4	17332.973	24	17334.684	44	17336.708	64	17338.624
5	17333.136	25	17334.777	45	17336.887	65	17338.727
6	17333.260	26	17334.940	46	17336.935	66	17338.915
7	17333.333	27	17335.025	47	17337.006	67	17338.957
8	17333.381	28	17335.252	48	17337.142	68	17339.028
9	17333.441	29	17335.354	49	17337.209		
10	17333.641	30	17335.408	50	17337.303		
11	17333.694	31	17335.483	51	17337.399		
12	17333.756	32	17335.632	52	17337.513		
13	17333.926	33	17335.714	53	17337.632		
14	17333.967	34	17335.768	54	17337.731		
15	17334.031	35	17335.879	55	17337.820		
16	17334.116	36	17335.968	56	17337.893		
17	17334.278	37	17336.042	57	17338.007		
18	17334.377	38	17336.134	58	17338.073		
19	17334.415	39	17336.241	59	17338.182		
20	17334.462	40	17336.335	60	17338.333		

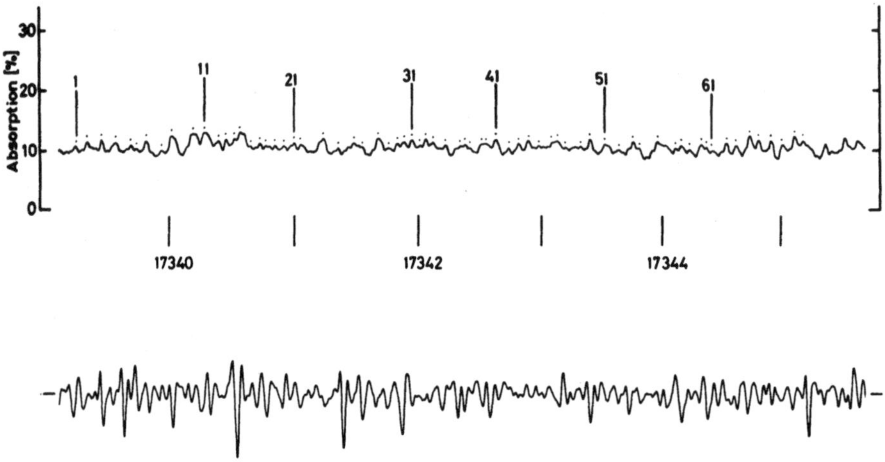

1	17339.253	21	17341.000	41	17342.627	61	17344.435
2	17339.344	22	17341.066	42	17342.725	62	17344.534
3	17339.462	23	17341.239	43	17342.807	63	17344.624
4	17339.569	24	17341.355	44	17342.889	64	17344.734
5	17339.688	25	17341.486	45	17342.985	65	17344.813
6	17339.753	26	17341.547	46	17343.109	66	17344.915
7	17339.817	27	17341.686	47	17343.146	67	17345.017
8	17339.942	28	17341.768	48	17343.201	68	17345.115
9	17340.025	29	17341.834	49	17343.333	69	17345.188
10	17340.187	30	17341.888	50	17343.400		
11	17340.277	31	17341.954	51	17343.515		
12	17340.389	32	17342.011	52	17343.623		
13	17340.458	33	17342.062	53	17343.753		
14	17340.520	34	17342.123	54	17343.804		
15	17340.581	35	17342.218	55	17343.953		
16	17340.658	36	17342.329	56	17344.104		
17	17340.732	37	17342.374	57	17344.151		
18	17340.782	38	17342.440	58	17344.223		
19	17340.855	39	17342.520	59	17344.316		
20	17340.924	40	17342.557	60	17344.365		

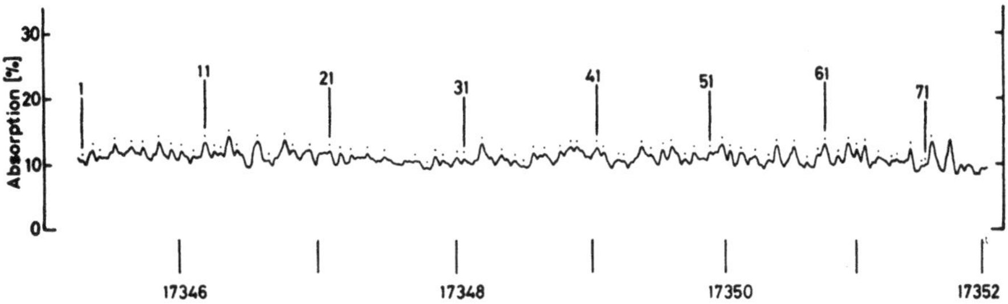

1	17345.310	21	17347.090	41	17349.036	61	17350.750
2	17345.373	22	17347.186	42	17349.088	62	17350.853
3	17345.437	23	17347.259	43	17349.194	63	17350.929
4	17345.545	24	17347.379	44	17349.234	64	17350.997
5	17345.664	25	17347.501	45	17349.373	65	17351.061
6	17345.752	26	17347.734	46	17349.432	66	17351.168
7	17345.867	27	17347.810	47	17349.524	67	17351.274
8	17345.957	28	17347.865	48	17349.595	68	17351.315
9	17346.024	29	17347.925	49	17349.694	69	17351.423
10	17346.118	30	17348.011	50	17349.755	70	17351.511
11	17346.200	31	17348.080	51	17349.873	71	17351.541
12	17346.255	32	17348.201	52	17349.964	72	17351.594
13	17346.310	33	17348.329	53	17350.026		
14	17346.363	34	17348.441	54	17350.105		
15	17346.427	35	17348.483	55	17350.212		
16	17346.562	36	17348.566	56	17350.315		
17	17346.640	37	17348.641	57	17350.380		
18	17346.769	38	17348.767	58	17350.514		
19	17346.833	39	17348.835	59	17350.615		
20	17346.953	40	17348.884	60	17350.697		

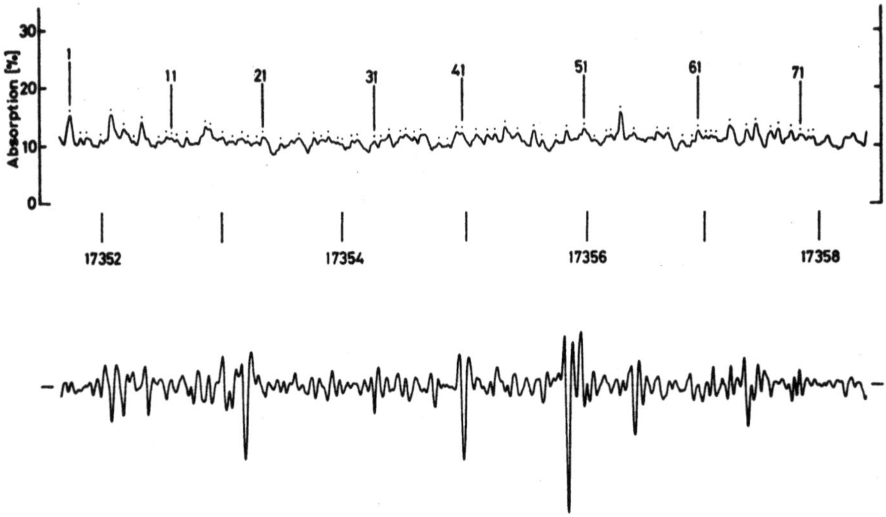

1	17351.727	21	17353.345	41	17354.989	61	17356.972
2	17351.814	22	17353.495	42	17355.109	62	17357.037
3	17351.861	23	17353.645	43	17355.203	63	17357.080
4	17351.977	24	17353.774	44	17355.263	64	17357.121
5	17352.067	25	17353.833	45	17355.346	65	17357.237
6	17352.178	26	17353.893	46	17355.451	66	17357.386
7	17352.261	27	17353.968	47	17355.586	67	17357.471
8	17352.331	28	17354.006	48	17355.655	68	17357.596
9	17352.473	29	17354.088	49	17355.768	69	17357.666
10	17352.539	30	17354.132	50	17355.859	70	17357.780
11	17352.580	31	17354.275	51	17356.000	71	17357.868
12	17352.624	32	17354.328	52	17356.099	72	17357.935
13	17352.703	33	17354.393	53	17356.187	73	17357.970
14	17352.860	34	17354.487	54	17356.229		
15	17352.900	35	17354.527	55	17356.313		
16	17353.009	36	17354.598	56	17356.427		
17	17353.093	37	17354.653	57	17356.633		
18	17353.174	38	17354.797	58	17356.716		
19	17353.221	39	17354.874	59	17356.837		
20	17353.294	40	17354.949	60	17356.926		

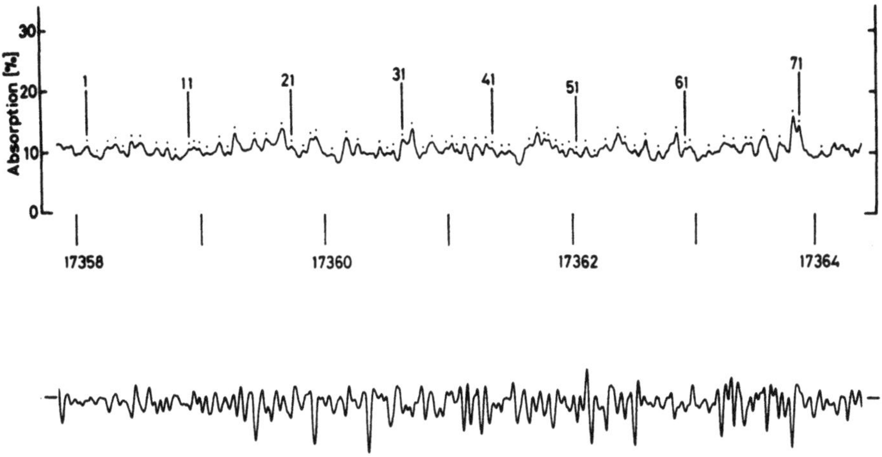

1	17358.096	21	17359.734	41	17361.359	61	17362.911
2	17358.165	22	17359.830	42	17361.427	62	17362.954
3	17358.258	23	17359.888	43	17361.483	63	17363.112
4	17358.318	24	17359.939	44	17361.651	64	17363.242
5	17358.387	25	17360.074	45	17361.717	65	17363.312
6	17358.446	26	17360.178	46	17361.768	66	17363.419
7	17358.520	27	17360.267	47	17361.803	67	17363.457
8	17358.651	28	17360.443	48	17361.862	68	17363.565
9	17358.732	29	17360.494	49	17361.912	69	17363.696
10	17358.802	30	17360.554	50	17361.987	70	17363.816
11	17358.910	31	17360.634	51	17362.049	71	17363.864
12	17358.945	32	17360.709	52	17362.106	72	17364.058
13	17358.989	33	17360.802	53	17362.164		
14	17359.050	34	17360.848	54	17362.264		
15	17359.160	35	17360.975	55	17362.365		
16	17359.233	36	17361.030	56	17362.413		
17	17359.289	37	17361.079	57	17362.502		
18	17359.441	38	17361.131	58	17362.589		
19	17359.540	39	17361.213	59	17362.694		
20	17359.662	40	17361.309	60	17362.841		

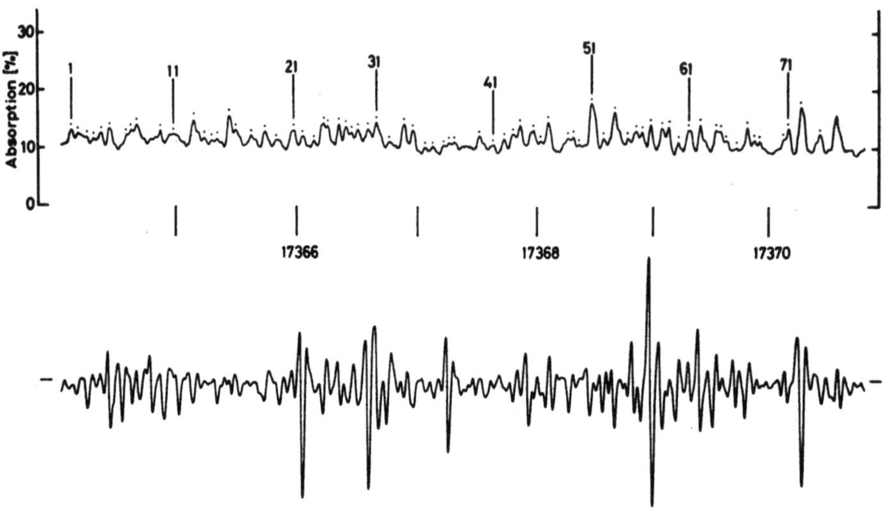

1	17364.161	21	17365.979	41	17367.633	61	17369.317
2	17364.224	22	17366.057	42	17367.728	62	17369.413
3	17364.280	23	17366.150	43	17367.807	63	17369.545
4	17364.331	24	17366.227	44	17367.862	64	17369.584
5	17364.392	25	17366.263	45	17367.972	65	17369.630
6	17364.467	26	17366.354	46	17368.035	66	17369.707
7	17364.609	27	17366.409	47	17368.100	67	17369.815
8	17364.647	28	17366.454	48	17368.273	68	17369.873
9	17364.692	29	17366.506	49	17368.307	69	17369.927
10	17364.879	30	17366.591	50	17368.363	70	17370.133
11	17364.977	31	17366.660	51	17368.467	71	17370.173
12	17365.158	32	17366.765	52	17368.569	72	17370.284
13	17365.252	33	17366.884	53	17368.665	73	17370.447
14	17365.307	34	17366.959	54	17368.785		
15	17365.357	35	17367.059	55	17368.843		
16	17365.458	36	17367.128	56	17368.903		
17	17365.513	37	17367.217	57	17368.981		
18	17365.639	38	17367.266	58	17369.073		
19	17365.746	39	17367.310	59	17369.136		
20	17365.836	40	17367.519	60	17369.220		

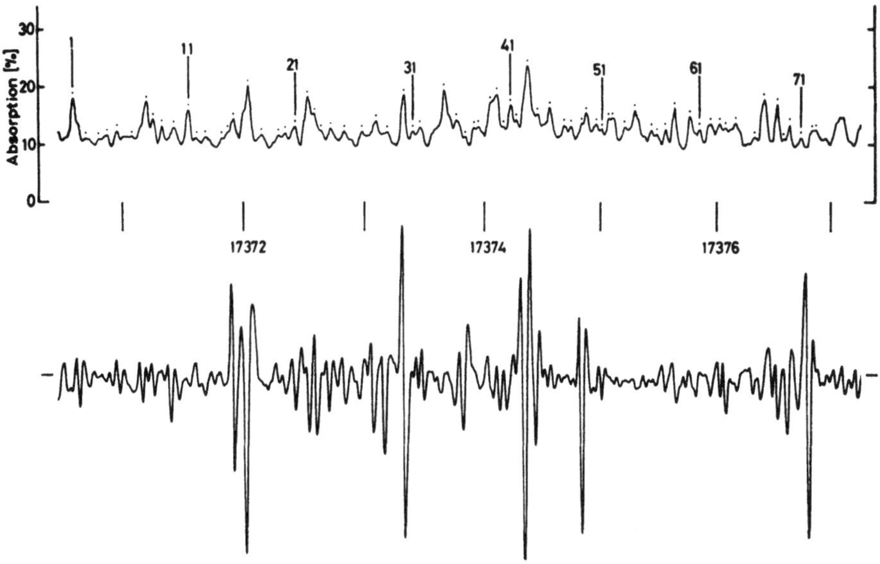

1	17370.588	21	17372.429	41	17374.230	61	17375.850
2	17370.703	22	17372.535	42	17374.280	62	17375.940
3	17370.808	23	17372.590	43	17374.376	63	17376.025
4	17370.865	24	17372.646	44	17374.466	64	17376.075
5	17370.953	25	17372.725	45	17374.564	65	17376.163
6	17371.074	26	17372.836	46	17374.695	66	17376.328
7	17371.199	27	17372.986	47	17374.752	67	17376.415
8	17371.254	28	17373.101	48	17374.840	68	17376.527
9	17371.327	29	17373.194	49	17374.880	69	17376.590
10	17371.426	30	17373.332	50	17374.968	70	17376.641
11	17371.546	31	17373.407	51	17375.014	71	17376.741
12	17371.612	32	17373.465	52	17375.074	72	17376.846
13	17371.685	33	17373.667	53	17375.109	73	17376.880
14	17371.816	34	17373.769	54	17375.217		
15	17371.867	35	17373.845	55	17375.297		
16	17371.917	36	17373.918	56	17375.433		
17	17372.039	37	17373.954	57	17375.485		
18	17372.150	38	17374.054	58	17375.558		
19	17372.293	39	17374.111	59	17375.638		
20	17372.348	40	17374.170	60	17375.769		

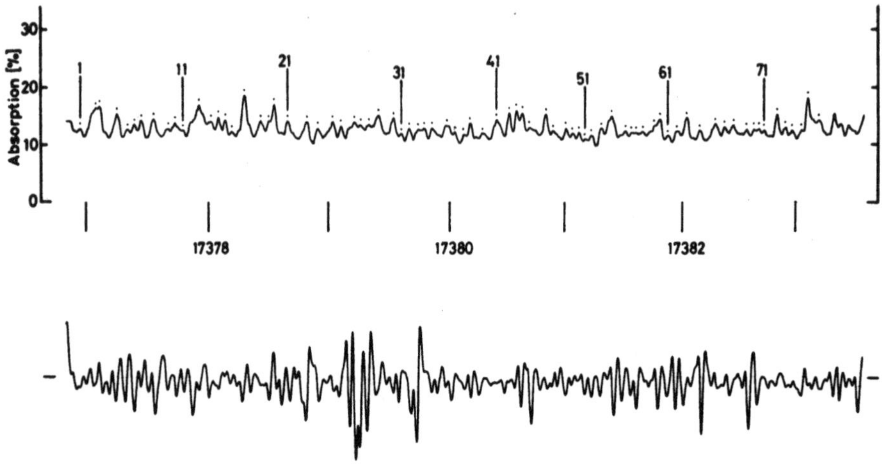

1	17376.960	21	17378.667	41	17380.412	61	17381.879
2	17377.093	22	17378.826	42	17380.520	62	17381.959
3	17377.123	23	17378.921	43	17380.581	63	17382.042
4	17377.275	24	17379.045	44	17380.633	64	17382.156
5	17377.355	25	17379.116	45	17380.705	65	17382.291
6	17377.419	26	17379.226	46	17380.839	66	17382.372
7	17377.468	27	17379.277	47	17380.900	67	17382.443
8	17377.569	28	17379.340	48	17381.016	68	17382.564
9	17377.683	29	17379.424	49	17381.072	69	17382.636
10	17377.735	30	17379.541	50	17381.126	70	17382.677
11	17377.802	31	17379.605	51	17381.183	71	17382.724
12	17377.933	32	17379.674	52	17381.230	72	17382.836
13	17378.031	33	17379.748	53	17381.310	73	17382.911
14	17378.100	34	17379.796	54	17381.401	74	17382.963
15	17378.153	35	17379.868	55	17381.522	75	17383.051
16	17378.218	36	17379.986	56	17381.571	76	17383.106
17	17378.315	37	17380.055	57	17381.609	77	17383.205
18	17378.451	38	17380.140	58	17381.667		
19	17378.516	39	17380.187	59	17381.770		
20	17378.560	40	17380.300	60	17381.811		

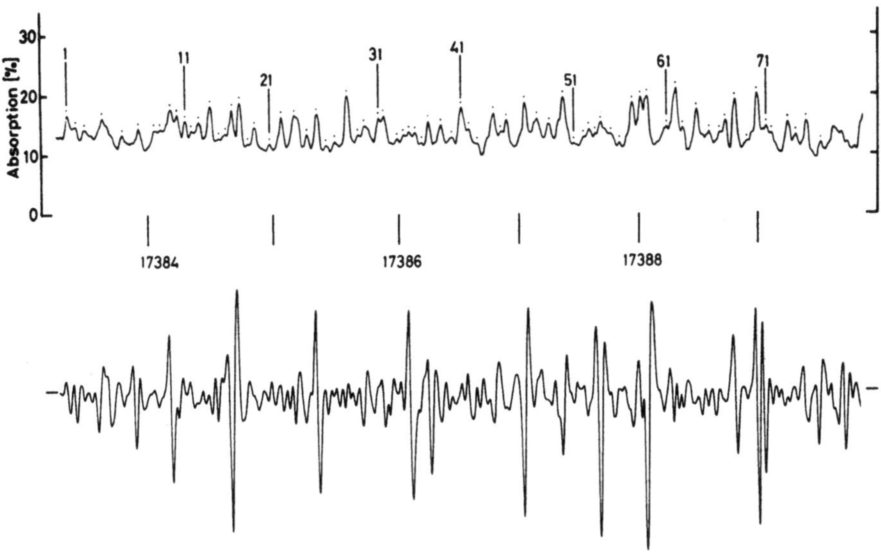

1	17383.340	21	17384.976	41	17386.529	61	17388.231
2	17383.407	22	17385.072	42	17386.590	62	17388.316
3	17383.477	23	17385.179	43	17386.661	63	17388.376
4	17383.623	24	17385.278	44	17386.791	64	17388.492
5	17383.784	25	17385.358	45	17386.852	65	17388.594
6	17383.918	26	17385.437	46	17386.903	66	17388.689
7	17384.044	27	17385.502	47	17387.055	67	17388.728
8	17384.092	28	17385.599	48	17387.160	68	17388.807
9	17384.175	29	17385.676	49	17387.256	69	17388.904
10	17384.230	30	17385.739	50	17387.378	70	17388.991
11	17384.294	31	17385.853	51	17387.469	71	17389.068
12	17384.346	32	17385.897	52	17387.549	72	17389.114
13	17394.405	33	17385.995	53	17387.598	73	17389.253
14	17384.495	34	17386.050	54	17387.647	74	17389.321
15	17384.565	35	17386.098	55	17387.696	75	17389.417
16	17384.620	36	17386.144	56	17387.785	76	17389.541
17	17384.669	37	17386.193	57	17387.853		
18	17384.732	38	17386.264	58	17387.956		
19	17384.796	39	17386.367	59	17388.026		
20	17384.852	40	17386.454	60	17388.076		

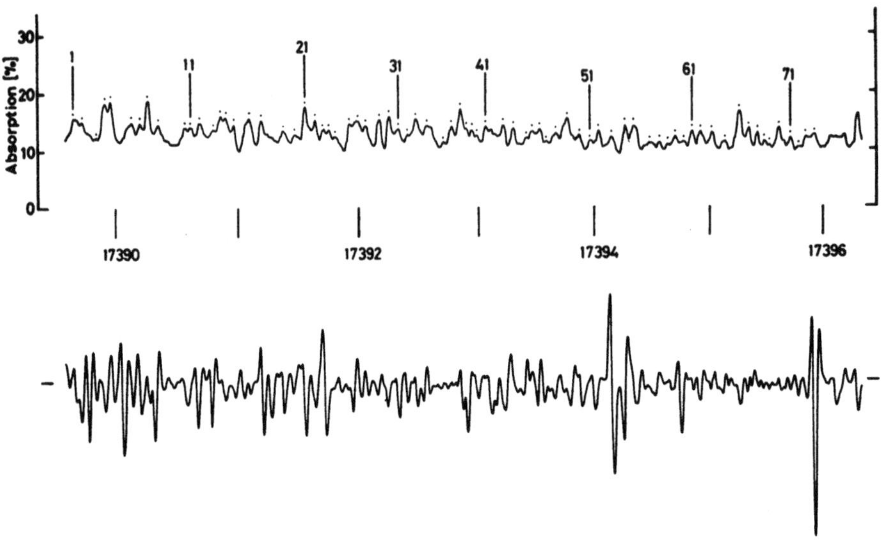

1	17389.647	21	17391.558	41	17393.077	61	17394.851
2	17389.718	22	17391.644	42	17393.222	62	17394.924
3	17389.833	23	17391.716	43	17393.308	63	17395.023
4	17389.910	24	17391.758	44	17393.414	64	17395.127
5	17389.962	25	17391.816	45	17393.473	65	17395.262
6	17390.126	26	17391.928	46	17393.536	66	17395.344
7	17390.200	27	17391.990	47	17393.604	67	17395.425
8	17390.265	28	17392.064	48	17393.680	68	17395.490
9	17390.346	29	17392.174	49	17393.769	69	17395.534
10	17390.562	30	17392.257	50	17393.877	70	17395.613
11	17390.611	31	17392.331	51	17393.969	71	17395.716
12	17390.680	32	17392.411	52	17394.042	72	17395.791
13	17390.800	33	17392.474	53	17394.154	73	17395.863
14	17390.862	34	17392.577	54	17394.274	74	17395.930
15	17390.905	35	17392.713	55	17394.346		
16	17390.972	36	17392.778	56	17394.505		
17	17391.098	37	17392.858	57	17394.570		
18	17391.195	38	17392.901	58	17394.647		
19	17391.372	39	17392.955	59	17394.701		
20	17391.472	40	17392.995	60	17394.777		

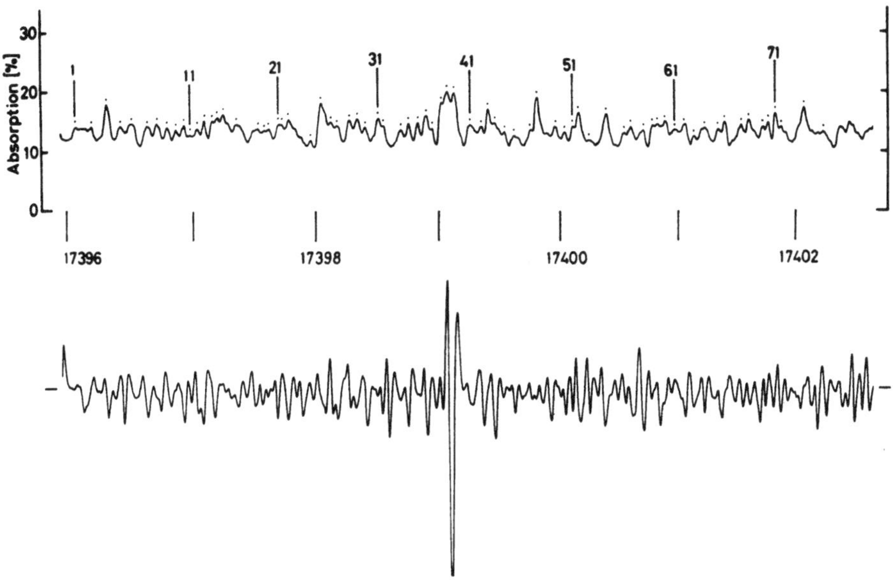

1	17396.076	21	17397.698	41	17399.261	61	17400.968
2	17396.196	22	17397.729	42	17399.355	62	17401.061
3	17396.320	23	17397.786	43	17399.410	63	17401.138
4	17396.432	24	17397.960	44	17399.466	64	17401.234
5	17396.507	25	17398.047	45	17399.549	65	17401.352
6	17396.640	26	17398.129	46	17399.621	66	17401.404
7	17396.721	27	17398.180	47	17399.760	67	17401.539
8	17396.797	28	17398.276	48	17399.814	68	17401.605
9	17396.872	29	17398.336	49	17399.967	69	17401.723
10	17396.933	30	17398.406	50	17400.039	70	17401.770
11	17396.980	31	17398.510	51	17400.113	71	17401.828
12	17397.043	32	17398.561	52	17400.166	72	17401.884
13	17397.100	33	17398.691	53	17400.249	73	17402.077
14	17397.164	34	17398.760	54	17400.398	74	17402.246
15	17397.203	35	17398.836	55	17400.545		
16	17397.247	36	17398.899	56	17400.606		
17	17397.347	37	17398.955	57	17400.707		
18	17397.529	38	17399.033	58	17400.793		
19	17397.582	39	17399.085	59	17400.837		
20	17397.625	40	17399.133	60	17400.900		

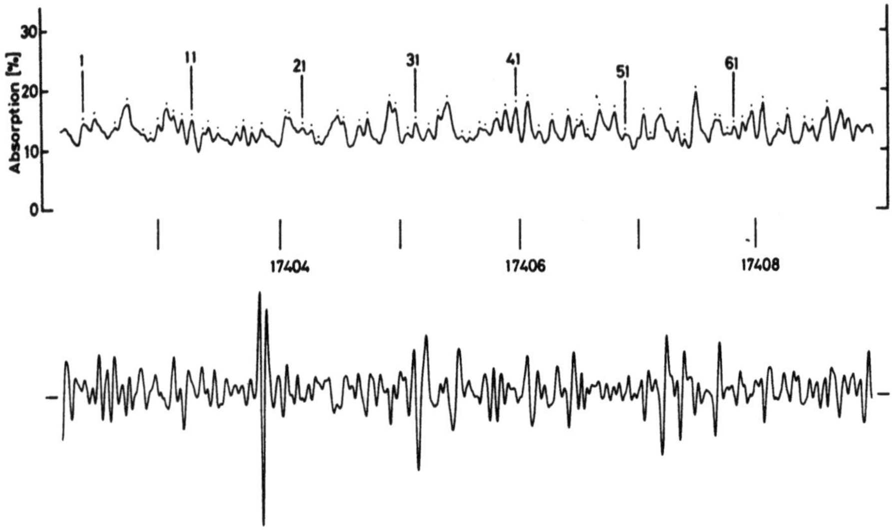

1	17402.397	21	17404.195	41	17405.976	61	17407.815
2	17402.485	22	17404.268	42	17406.074	62	17407.892
3	17402.660	23	17404.326	43	17406.175	63	17407.973
4	17402.766	24	17404.487	44	17406.277	64	17408.070
5	17402.892	25	17404.531	45	17406.418	65	17408.208
6	17402.947	26	17404.671	46	17406.487	66	17408.284
7	17403.013	27	17404.737	47	17406.528	67	17408.433
8	17403.078	28	17404.919	48	17406.576	68	17408.498
9	17403.136	29	17404.963	49	17406.682	69	17408.628
10	17403.205	30	17405.065	50	17406.803		
11	17403.287	31	17405.137	51	17406.902		
12	17403.375	32	17405.247	52	17407.053		
13	17403.427	33	17405.331	53	17407.110		
14	17403.499	34	17405.404	54	17407.196		
15	17403.651	35	17405.535	55	17407.336		
16	17403.711	36	17405.600	56	17407.396		
17	17403.782	37	17405.681	57	17407.492		
18	17403.858	38	17405.724	58	17407.657		
19	17404.050	39	17405.818	59	17407.699		
20	17404.086	40	17405.890	60	17407.762		

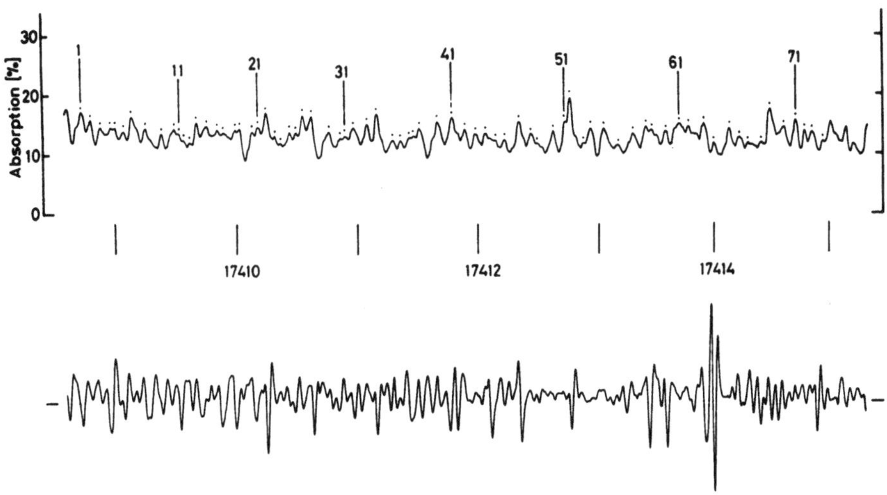

1	17408.743	21	17410.182	41	17411.785	61	17413.697
2	17408.813	22	17410.252	42	17411.891	62	17413.774
3	17408.891	23	17410.316	43	17411.980	63	17413.912
4	17408.975	24	17410.373	44	17412.062	64	17414.002
5	17409.009	25	17410.449	45	17412.146	65	17414.132
6	17409.081	26	17410.499	46	17412.219	66	17414.235
7	17409.151	27	17410.553	47	17412.335	67	17414.297
8	17409.258	28	17410.626	48	17412.435	68	17414.422
9	17409.389	29	17410.775	49	17412.488	69	17414.492
10	17409.499	30	17410.869	50	17412.624	70	17414.612
11	17409.537	31	17410.906	51	17412.719	71	17414.710
12	17409.574	32	17410.975	52	17412.766	72	17414.790
13	17409.625	33	17411.086	53	17412.876	73	17414.858
14	17409.685	34	17411.164	54	17412.940	74	17414.952
15	17409.764	35	17411.299	55	17413.050		
16	17409.852	36	17411.363	56	17413.172		
17	17409.909	37	17411.433	57	17413.298		
18	17409.993	38	17411.467	58	17413.411		
19	17410.027	39	17411.517	59	17413.458		
20	17410.133	40	17411.666	60	17413.585		

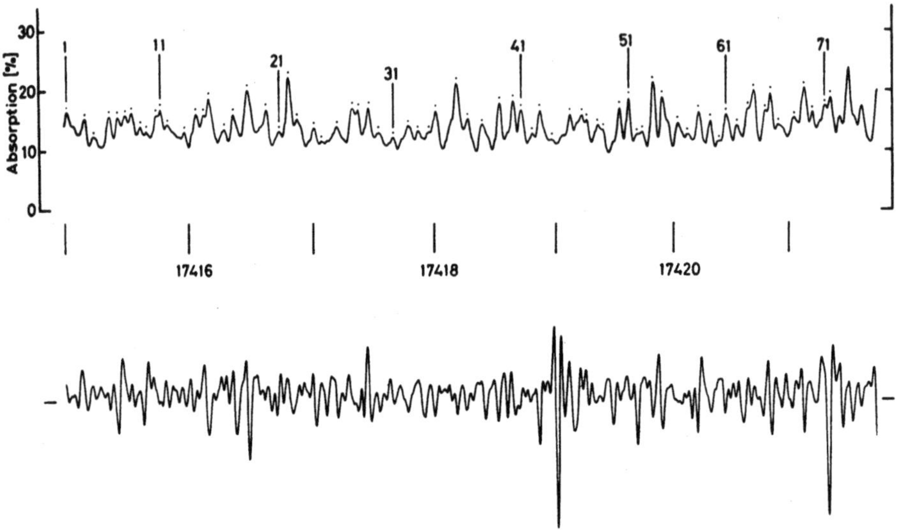

1	17415.021	21	17416.727	41	17418.723	61	17420.443
2	17415.154	22	17416.808	42	17418.820	62	17420.542
3	17415.221	23	17416.876	43	17418.878	63	17420.631
4	17415.348	24	17417.015	44	17418.978	64	17420.691
5	17415.419	25	17417.079	45	17419.121	65	17420.787
6	17415.481	26	17417.200	46	17419.221	66	17420.836
7	17415.533	27	17417.337	47	17419.261	67	17420.902
8	17415.604	28	17417.376	48	17419.351	68	17421.043
9	17415.648	29	17417.461	49	17419.393	69	17421.134
10	17415.733	30	17417.546	50	17419.533	70	17421.205
11	17415.763	31	17417.660	51	17419.611	71	17421.313
12	17415.825	32	17417.801	52	17419.684	72	17421.353
13	17415.960	33	17417.866	53	17419.728		
14	17416.060	34	17417.957	54	17419.822		
15	17416.115	35	17418.015	55	17419.903		
16	17416.163	36	17418.187	56	17420.038		
17	17416.293	37	17418.279	57	17420.123		
18	17416.366	38	17418.402	58	17420.212		
19	17416.482	39	17418.545	59	17420.313		
20	17416.625	40	17418.656	60	17420.387		

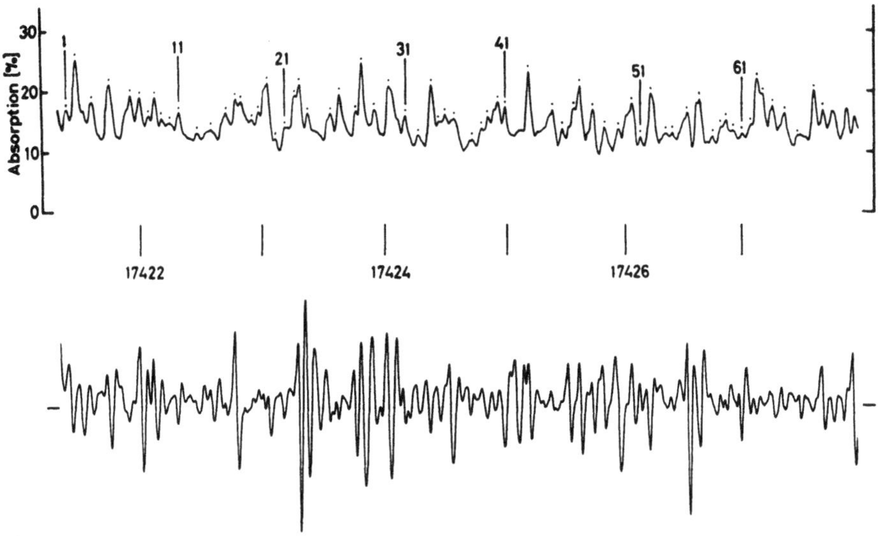

1	17421.436	21	17423.188	41	17424.997	61	17427.010
2	17421.514	22	17423.277	42	17425.191	62	17427.085
3	17421.628	23	17423.311	43	17425.391	63	17427.145
4	17421.766	24	17423.377	44	17425.472	64	17427.191
5	17421.936	25	17423.565	45	17425.572	65	17427.274
6	17422.008	26	17423.640	46	17425.621	66	17427.381
7	17422.080	27	17423.772	47	17425.732	67	17427.496
8	17422.134	28	17423.821	48	17425.827	68	17427.629
9	17422.187	29	17423.923	49	17425.949	69	17427.706
10	17422.251	30	17424.041	50	17426.062		
11	17422.328	31	17424.176	51	17426.136		
12	17422.476	32	17424.280	52	17426.227		
13	17422.584	33	17424.389	53	17426.334		
14	17422.704	34	17424.449	54	17426.412		
15	17422.786	35	17424.500	55	17426.543		
16	17422.827	36	17424.574	56	17426.650		
17	17422.918	37	17424.723	57	17426.759		
18	17422.975	38	17424.802	58	17426.840		
19	17423.049	39	17424.854	59	17426.879		
20	17423.115	40	17424.936	60	17426.932		

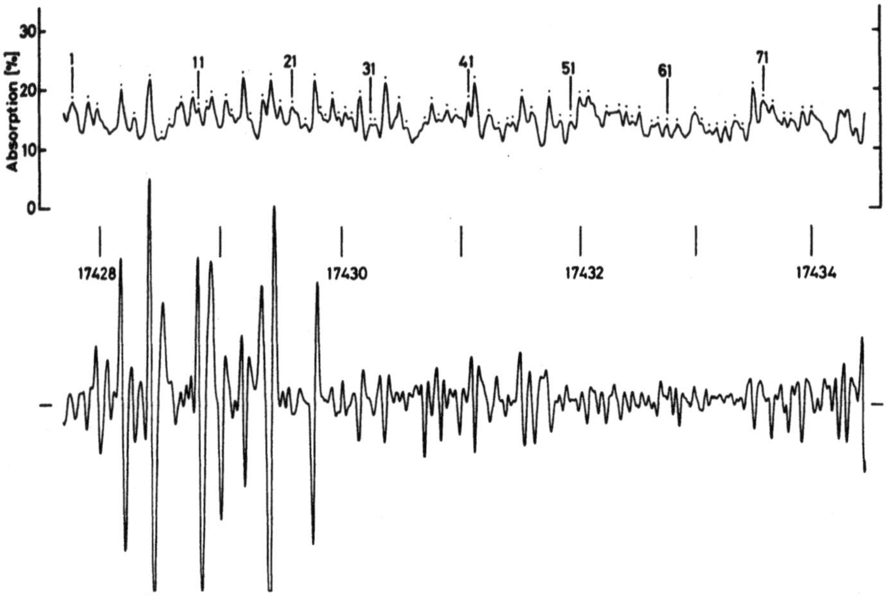

1	17427.786	21	17429.591	41	17431.065	61	17432.740
2	17427.914	22	17429.704	42	17431.116	62	17432.839
3	17427.986	23	17429.784	43	17431.237	63	17432.992
4	17428.190	24	17429.832	44	17431.307	64	17433.047
5	17428.292	25	17429.868	45	17431.387	65	17433.109
6	17428.422	26	17429.932	46	17431.432	66	17433.172
7	17428.522	27	17429.986	47	17431.510	67	17433.249
8	17428.582	28	17430.036	48	17431.593	68	17433.331
9	17428.686	29	17430.092	49	17431.733	69	17433.410
10	17428.779	30	17430.157	50	17431.822	70	17433.501
11	17428.829	31	17430.241	51	17431.914	71	17433.588
12	17428.899	32	17430.282	52	17431.992	72	17433.674
13	17428.939	33	17430.372	53	17432.065	73	17433.772
14	17429.057	34	17430.484	54	17432.224	74	17433.832
15	17429.101	35	17430.543	55	17432.327	75	17433.929
16	17429.196	36	17430.707	56	17432.389	76	17433.999
17	17429.255	37	17430.760	57	17432.446		
18	17429.352	38	17430.824	58	17432.508		
19	17429.415	39	17430.888	59	17432.615		
20	17429.494	40	17430.957	60	17432.669		

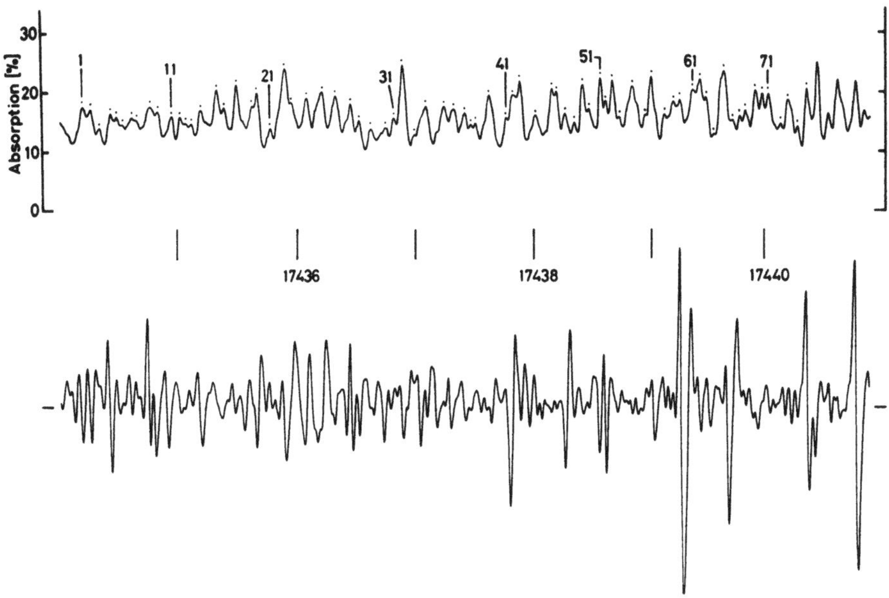

1	17434.255	21	17435.775	41	17437.777	61	17439.378
2	17434.316	22	17435.897	42	17437.825	62	17439.438
3	17434.384	23	17435.948	43	17437.895	63	17439.496
4	17434.475	24	17436.080	44	17438.025	64	17439.561
5	17434.519	25	17436.213	45	17438.165	65	17439.652
6	17434.568	26	17436.327	46	17438.205	66	17439.734
7	17434.641	27	17436.457	47	17438.280	67	17439.797
8	17434.680	28	17436.528	48	17438.350	68	17439.843
9	17434.788	29	17436.627	49	17438.425	69	17439.926
10	17434.848	30	17436.751	50	17438.479	70	17439.991
11	17434.962	31	17436.820	51	17438.574	71	17440.044
12	17435.030	32	17436.897	52	17438.619	72	17440.153
13	17435.073	33	17437.000	53	17438.678	73	17440.220
14	17435.123	34	17437.094	54	17438.738	74	17440.307
15	17435.198	35	17437.246	55	17438.850	75	17440.386
16	17435.326	36	17437.329	56	17438.962		
17	17435.387	37	17437.424	57	17439.014		
18	17435.490	38	17437.474	58	17439.125		
19	17435.620	39	17437.519	59	17439.211		
20	17435.662	40	17437.632	60	17439.261		

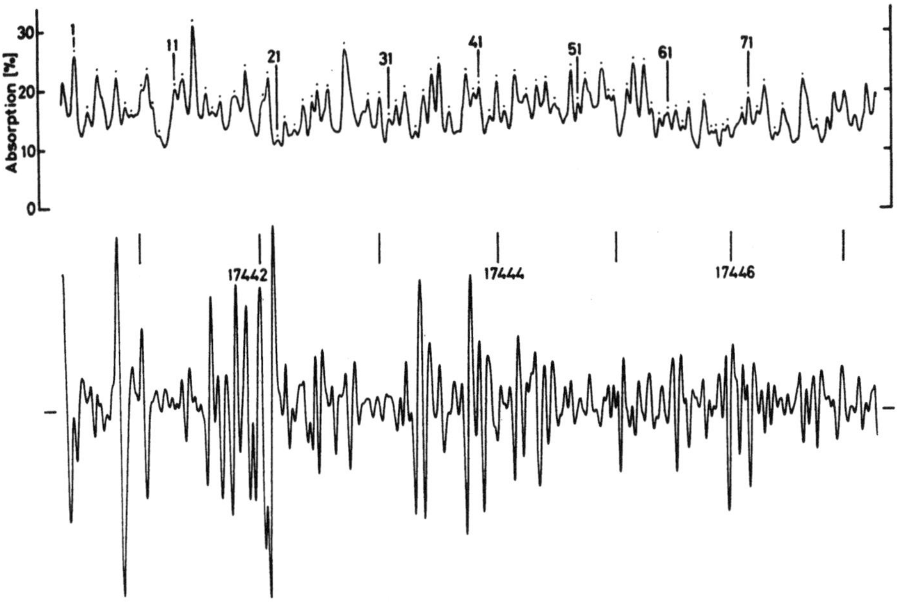

1	17440.481	21	17442.164	41	17443.849	61	17445.451
2	17440.582	22	17442.224	42	17443.932	62	17445.524
3	17440.662	23	17442.300	43	17443.996	63	17445.585
4	17440.820	24	17442.380	44	17444.058	64	17445.637
5	17440.892	25	17442.454	45	17444.151	65	17445.775
6	17440.942	26	17442.497	46	17444.244	66	17445.838
7	17441.035	27	17442.590	47	17444.334	67	17445.876
8	17441.078	28	17442.737	48	17444.411	68	17445.942
9	17441.123	29	17442.920	49	17444.482	69	17445.982
10	17441.176	30	17443.014	50	17444.625	70	17446.117
11	17441.309	31	17443.093	51	17444.687	71	17446.174
12	17441.369	32	17443.151	52	17444.754	72	17446.251
13	17441.463	33	17443.221	53	17444.896	73	17446.317
14	17441.566	34	17443.302	54	17444.952	74	17446.411
15	17441.621	35	17443.378	55	17444.989	75	17446.482
16	17441.684	36	17443.451	56	17445.109	76	17446.659
17	17441.803	37	17443.516	57	17445.167	77	17446.774
18	17441.900	38	17443.595	58	17445.256		
19	17442.040	39	17443.741	59	17445.316		
20	17442.086	40	17443.801	60	17445.386		

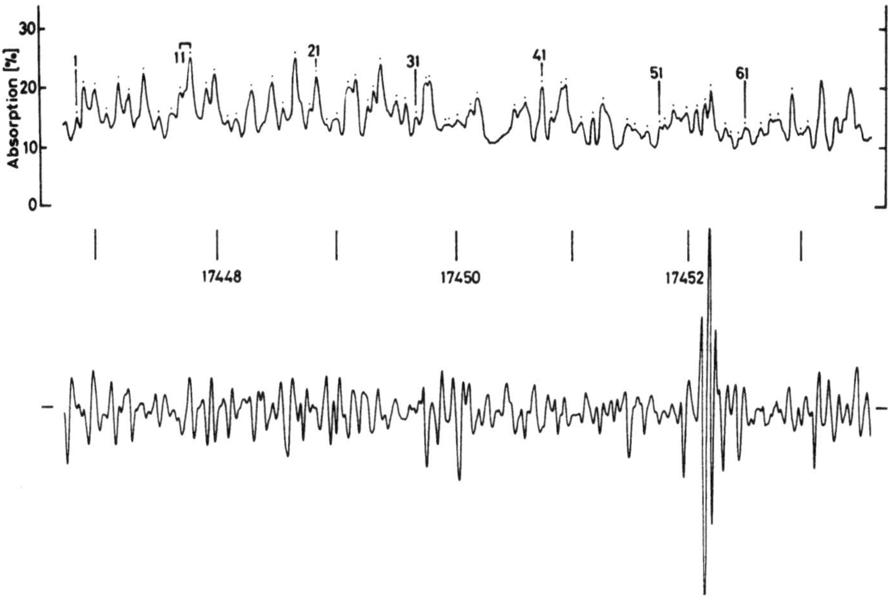

1	17446.868	21	17448.845	41	17450.746	61	17452.515
2	17446.928	22	17448.929	42	17450.822	62	17452.636
3	17447.015	23	17449.006	43	17450.917	63	17452.730
4	17447.110	24	17449.112	44	17450.953	64	17452.784
5	17447.211	25	17449.171	45	17451.081	65	17452.925
6	17447.289	26	17449.267	46	17451.183	66	17453.003
7	17447.416	27	17449.315	47	17451.267	67	17453.065
8	17447.535	28	17449.379	48	17451.478		
9	17447.636	29	17449.506	49	17451.546		
10	17447.711	30	17449.587	50	17451.639		
11	17447.796	31	17449.675	51	17451.752		
12	17447.923	32	17449.761	52	17451.794		
13	17447.988	33	17449.800	53	17451.876		
14	17448.095	34	17449.926	54	17451.986		
15	17448.169	35	17449.962	55	17452.078		
16	17448.298	36	17450.020	56	17452.152		
17	17448.474	37	17450.131	57	17452.208		
18	17448.563	38	17450.193	58	17452.327		
19	17448.669	39	17450.502	59	17452.373		
20	17448.791	40	17450.602	60	17452.454		

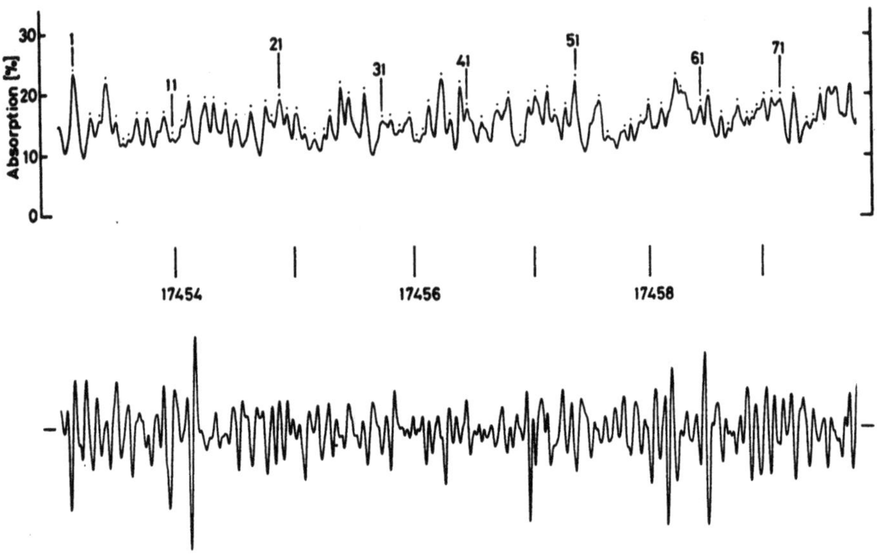

1	17453.189	21	17454.888	41	17456.451	61	17458.454
2	17453.314	22	17454.951	42	17456.567	62	17458.518
3	17453.384	23	17455.026	43	17456.699	63	17458.632
4	17453.448	24	17455.099	44	17456.789	64	17458.698
5	17453.529	25	17455.180	45	17456.900	65	17458.783
6	17453.585	26	17455.264	46	17456.962	66	17458.869
7	17453.635	27	17455.311	47	17457.017	67	17458.911
8	17453.703	28	17455.401	48	17457.121	68	17458.958
9	17453.789	29	17455.466	49	17457.189	69	17459.015
10	17453.926	30	17455.597	50	17457.279	70	17459.088
11	17453.995	31	17455.746	51	17457.361	71	17459.163
12	17454.068	32	17455.811	52	17457.574	72	17459.281
13	17454.126	33	17455.888	53	17457.651	73	17459.388
14	17454.259	34	17455.964	54	17457.793	74	17459.448
15	17454.332	35	17456.030	55	17457.846	75	17459.511
16	17454.433	36	17456.071	56	17457.933		
17	17454.521	37	17456.131	57	17457.998		
18	17454.650	38	17456.237	58	17458.110		
19	17454.772	39	17456.307	59	17458.239		
20	17454.821	40	17456.390	60	17458.298		

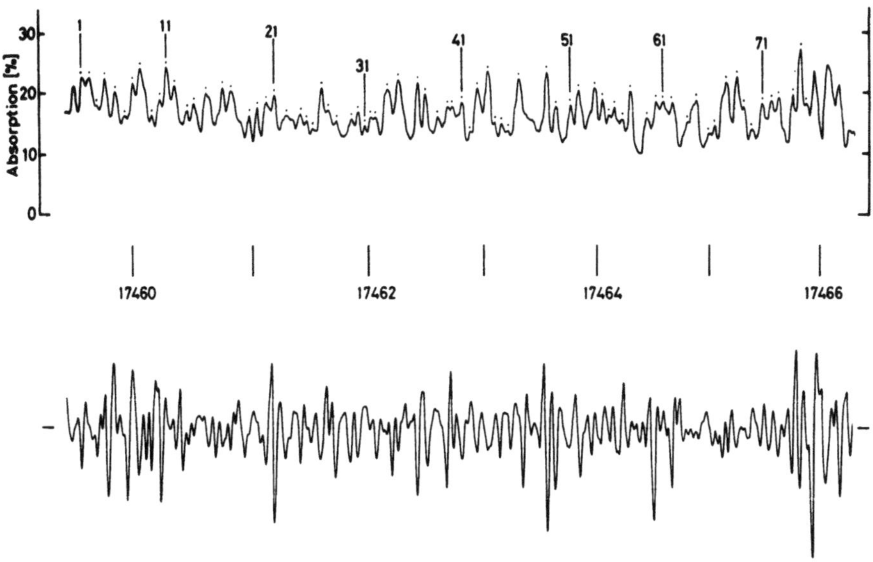

1	17459.583	21	17461.198	41	17462.815	61	17464.588
2	17459.643	22	17461.294	42	17462.875	62	17464.673
3	17459.706	23	17461.413	43	17462.951	63	17464.812
4	17459.776	24	17461.478	44	17463.043	64	17464.891
5	17459.861	25	17461.530	45	17463.098	65	17465.008
6	17459.936	26	17461.601	46	17463.151	66	17465.057
7	17460.012	27	17461.658	47	17463.216	67	17465.161
8	17460.080	28	17461.727	48	17463.306	68	17465.256
9	17460.172	29	17461.850	49	17463.557	69	17465.314
10	17460.241	30	17461.909	50	17463.631	70	17465.380
11	17460.299	31	17461.962	51	17463.770	71	17465.469
12	17460.362	32	17462.011	52	17463.838	72	17465.553
13	17460.464	33	17462.053	53	17463.981	73	17465.618
14	17460.520	34	17462.160	54	17464.048	74	17465.754
15	17460.621	35	17462.258	55	17464.111	75	17465.821
16	17460.760	36	17462.431	56	17464.151	76	17465.877
17	17460.827	37	17462.495	57	17464.222		
18	17460.979	38	17462.602	58	17464.295		
19	17461.049	39	17462.686	59	17464.438		
20	17461.128	40	17462.730	60	17464.520		

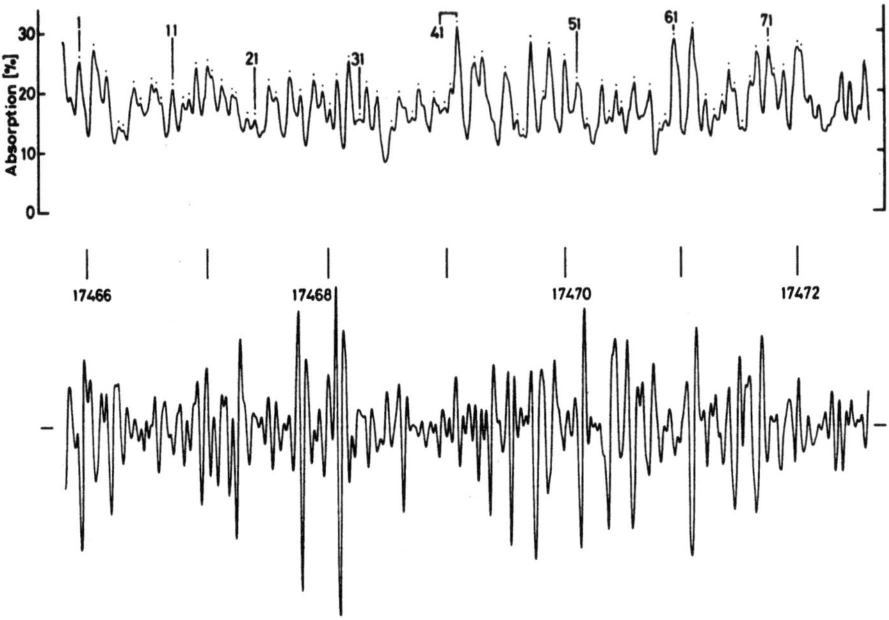

1	17465.949	21	17467.408	41	17469.092	61	17470.946
2	17466.066	22	17467.521	42	17469.238	62	17471.109
3	17466.179	23	17467.577	43	17469.299	63	17471.220
4	17466.279	24	17467.691	44	17469.491	64	17471.304
5	17466.320	25	17467.777	45	17469.588	65	17471.363
6	17466.410	26	17467.891	46	17469.636	66	17471.422
7	17466.464	27	17467.961	47	17469.705	67	17471.473
8	17466.554	28	17468.026	48	17469.812	68	17471.542
9	17466.592	29	17468.090	49	17469.867	69	17471.614
10	17466.637	30	17468.187	50	17469.997	70	17471.665
11	17466.730	31	17468.279	51	17470.108	71	17471.773
12	17466.813	32	17468.329	52	17470.198	72	17471.818
13	17466.863	33	17468.415	53	17470.325	73	17471.897
14	17466.924	34	17468.536	54	17470.389	74	17472.010
15	17467.015	35	17468.593	55	17470.449	75	17472.048
16	17467.052	36	17468.706	56	17470.500		
17	17467.137	37	17468.760	57	17470.604		
18	17467.220	38	17468.894	58	17470.741		
19	17467.249	39	17468.984	59	17470.822		
20	17467.344	40	17469.035	60	17470.873		

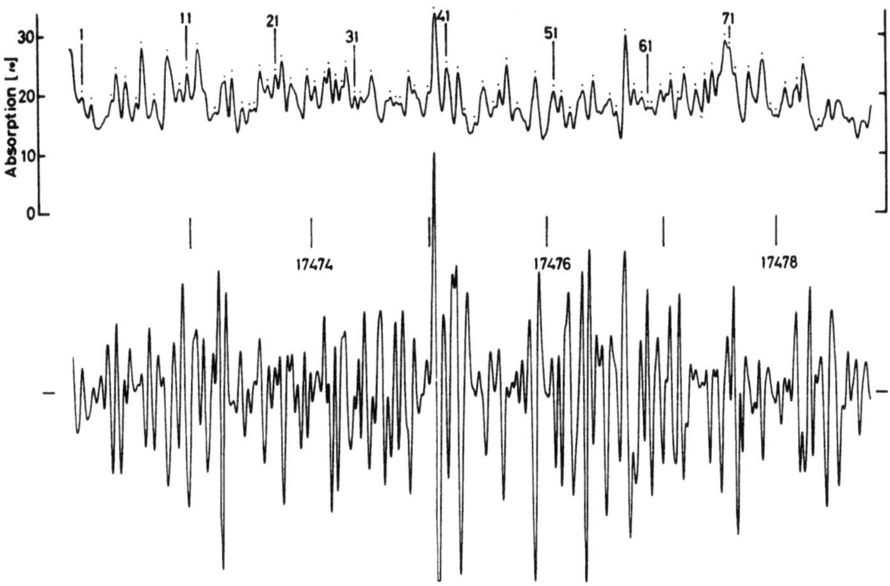

1	17472.135	21	17473.722	41	17475.155	61	17476.878
2	17472.209	22	17473.777	42	17475.254	62	17476.906
3	17472.372	23	17473.848	43	17475.306	63	17476.982
4	17472.413	24	17473.986	44	17475.390	64	17477.073
5	17472.487	25	17474.048	45	17475.471	65	17477.136
6	17472.569	26	17474.132	46	17475.550	66	17477.193
7	17472.624	27	17474.172	47	17475.602	67	17477.297
8	17472.722	28	17474.228	48	17475.663	68	17477.380
9	17472.834	29	17474.277	49	17475.744	69	17477.444
10	17472.929	30	17474.317	50	17475.912	70	17477.560
11	17472.988	31	17474.388	51	17476.061	71	17477.601
12	17473.077	32	17474.433	52	17476.127	72	17477.648
13	17473.209	33	17474.526	53	17476.202	73	17477.771
14	17473.289	34	17474.684	54	17476.312	74	17477.890
15	17473.352	35	17474.729	55	17476.397	75	17477.955
16	17473.436	36	17474.766	56	17476.543	76	17478.001
17	17473.503	37	17474.839	57	17476.602	77	17478.097
18	17473.540	38	17474.884	58	17476.689	78	17478.206
19	17473.590	39	17474.999	59	17476.753	79	17478.264
20	17473.651	40	17475.062	60	17476.823	80	17478.397

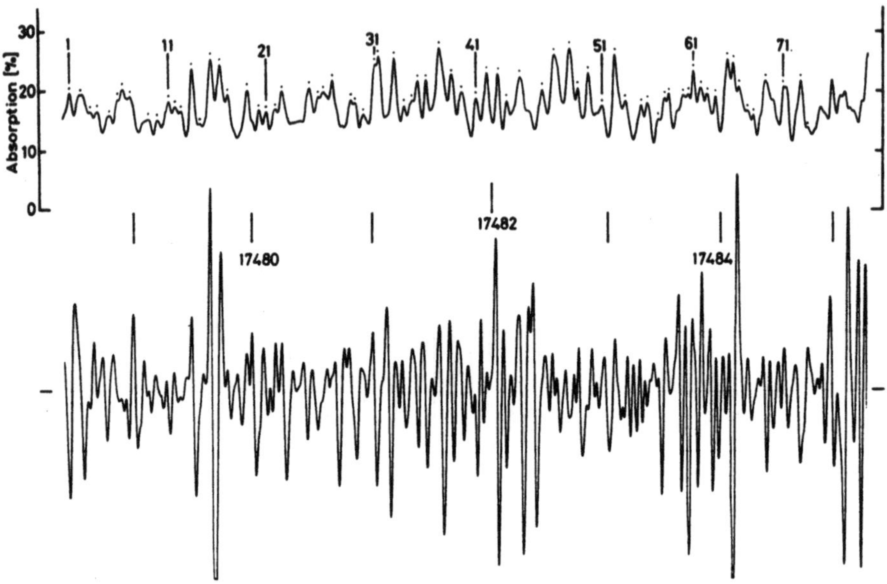

1	17478.487	21	17480.116	41	17481.877	61	17483.765
2	17478.577	22	17480.195	42	17481.967	62	17483.835
3	17478.648	23	17480.254	43	17482.062	63	17483.892
4	17478.695	24	17480.482	44	17482.137	64	17483.965
5	17478.806	25	17480.558	45	17482.249	65	17484.070
6	17478.869	26	17480.596	46	17482.440	66	17484.118
7	17478.918	27	17480.681	47	17482.542	67	17484.169
8	17478.974	28	17480.830	48	17482.681	68	17484.250
9	17479.117	29	17480.872	49	17482.754	69	17484.296
10	17479.203	30	17480.932	50	17482.845	70	17484.405
11	17479.297	31	17481.029	51	17482.958	71	17484.576
12	17479.351	32	17481.064	52	17483.074	72	17484.727
13	17479.398	33	17481.189	53	17483.162	73	17484.787
14	17479.491	34	17481.273	54	17483.302		
15	17479.550	35	17481.340	55	17483.361		
16	17479.650	36	17481.386	56	17483.448		
17	17479.726	37	17481.460	57	17483.506		
18	17479.789	38	17481.576	58	17483.545		
19	17479.963	39	17481.673	59	17483.674		
20	17480.059	40	17481.758	60	17483.721		

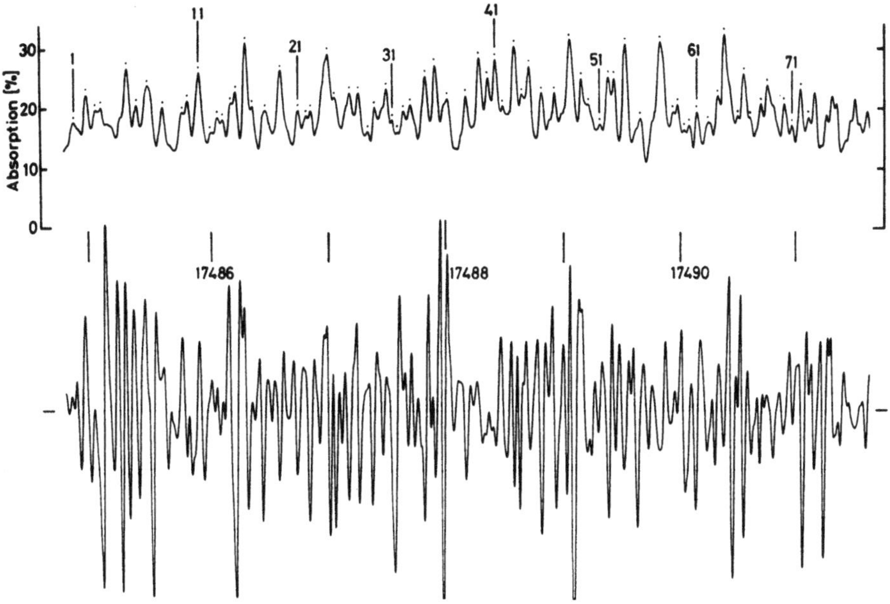

1	17484.891	21	17486.737	41	17488.419	61	17490.155
2	17484.995	22	17486.805	42	17488.481	62	17490.276
3	17485.068	23	17486.847	43	17488.578	63	17490.347
4	17485.110	24	17486.992	44	17488.646	64	17490.407
5	17485.322	25	17487.046	45	17488.701	65	17490.524
6	17485.394	26	17487.173	46	17488.807	66	17490.582
7	17485.485	27	17487.246	47	17488.920	67	17490.639
8	17485.610	28	17487.328	48	17489.055	68	17490.733
9	17485.767	29	17487.386	49	17489.153	69	17490.790
10	17485.815	30	17487.489	50	17489.204	70	17490.941
11	17485.913	31	17487.536	51	17489.315	71	17491.013
12	17486.002	32	17487.576	52	17489.386	72	17491.095
13	17486.064	33	17487.636	53	17489.438	73	17491.170
14	17486.099	34	17487.691	54	17489.536		
15	17486.175	35	17487.825	55	17489.665		
16	17486.215	36	17487.911	56	17489.842		
17	17486.299	37	17488.010	57	17489.945		
18	17486.347	38	17488.168	58	17489.981		
19	17486.451	39	17488.283	59	17490.042		
20	17486.592	40	17488.355	60	17490.085		

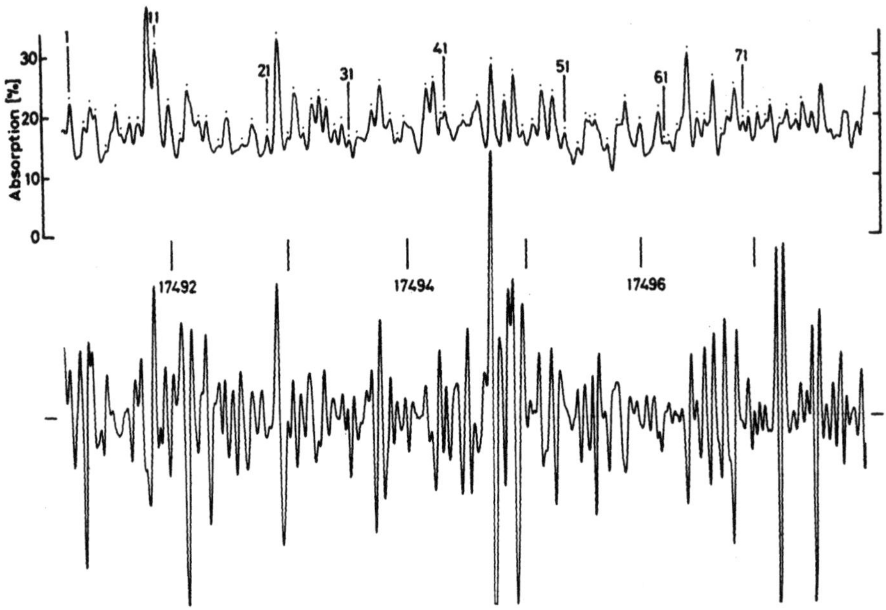

1	17491.225	21	17492.862	41	17494.358	61	17496.260
2	17491.336	22	17492.951	42	17494.510	62	17496.297
3	17491.385	23	17493.042	43	17494.633	63	17496.370
4	17491.427	24	17493.102	44	17494.748	64	17496.458
5	17491.509	25	17493.172	45	17494.865	65	17496.546
6	17491.599	26	17493.248	46	17494.943	66	17496.613
7	17491.643	27	17493.309	47	17495.019	67	17496.690
8	17491.713	28	17493.366	48	17495.107	68	17496.763
9	17491.781	29	17493.436	49	17495.182	69	17496.813
10	17491.865	30	17493.498	50	17495.274	70	17496.874
11	17491.930	31	17493.556	51	17495.378	71	17496.950
12	17492.030	32	17493.629	52	17495.495	72	17496.998
13	17492.127	33	17493.748	53	17495.570	73	17497.079
14	17492.185	34	17493.821	54	17495.601	74	17497.152
15	17492.279	35	17493.903	55	17495.644	75	17497.194
16	17492.346	36	17493.963	56	17495.750	76	17497.285
17	17492.451	37	17494.021	57	17495.843	77	17497.337
18	17492.522	38	17494.205	58	17495.909	78	17497.419
19	17492.654	39	17494.261	59	17496.041	79	17497.483
20	17492.739	40	17494.325	60	17496.207		

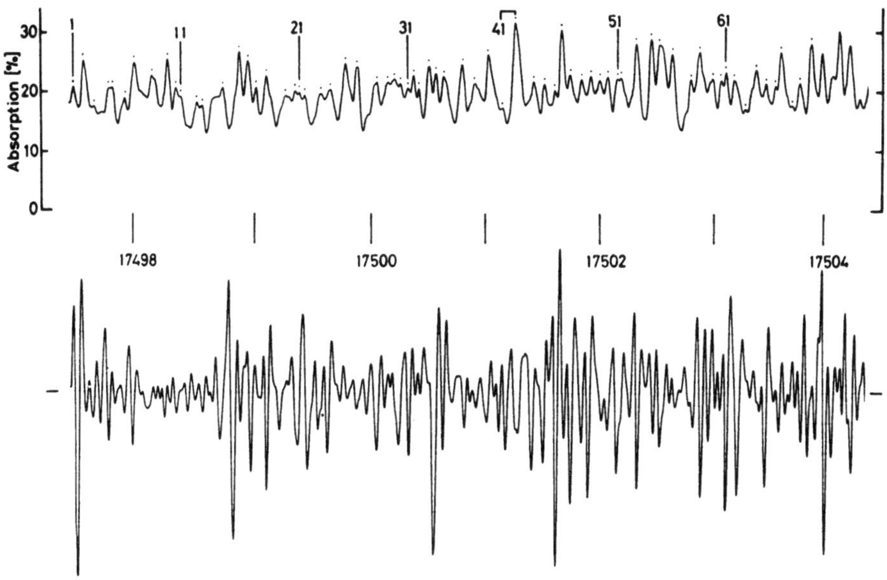

1	17497.561	21	17499.416	41	17501.268	61	17503.116
2	17497.647	22	17499.462	42	17501.421	62	17503.190
3	17497.729	23	17499.599	43	17501.514	63	17503.281
4	17497.851	24	17499.680	44	17501.610	64	17503.380
5	17497.879	25	17499.808	45	17501.675	65	17503.468
6	17497.974	26	17499.905	46	17501.744	66	17503.556
7	17498.054	27	17500.073	47	17501.846	67	17503.604
8	17498.197	28	17500.169	48	17501.929	68	17503.709
9	17498.329	29	17500.230	49	17502.007	69	17503.784
10	17498.397	30	17500.277	50	17502.074	70	17503.888
11	17498.437	31	17500.341	51	17502.159		
12	17498.566	32	17500.308	52	17502.191		
13	17498.609	33	17500.434	53	17502.329		
14	17498.780	34	17500.520	54	17502.457		
15	17498.918	35	17500.579	55	17502.529		
16	17498.990	36	17500.646	56	17502.637		
17	17499.059	37	17500.808	57	17502.798		
18	17499.143	38	17500.914	58	17502.878		
19	17499.301	39	17501.034	59	17502.991		
20	17499.380	40	17501.151	60	17503.070		

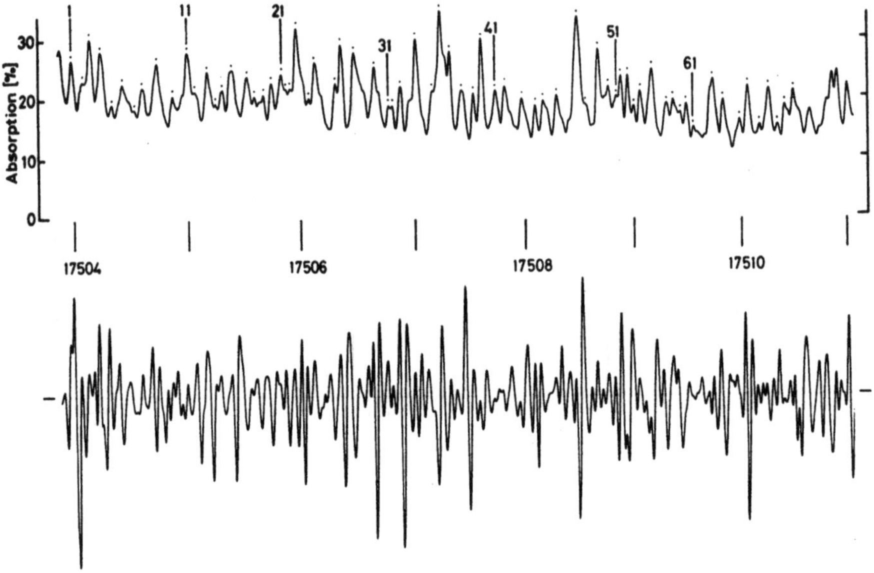

1	17503.989	21	17505.834	41	17507.729	61	17509.557
2	17504.080	22	17505.869	42	17507.819	62	17509.738
3	17504.148	23	17505.915	43	17507.974	63	17509.835
4	17504.241	24	17505.971	44	17508.092	64	17509.984
5	17504.338	25	17506.077	45	17508.167	65	17510.060
6	17504.432	26	17506.132	46	17508.295	66	17510.168
7	17504.526	27	17506.312	47	17508.484	67	17510.260
8	17504.605	28	17506.367	48	17508.570	68	17510.343
9	17504.727	29	17506.481	49	17508.677	69	17510.412
10	17504.867	30	17506.657	50	17508.769	70	17510.493
11	17504.997	31	17506.783	51	17508.844		
12	17505.063	32	17506.812	52	17508.882		
13	17505.175	33	17506.885	53	17508.949		
14	17505.255	34	17507.022	54	17509.002		
15	17505.309	35	17507.163	55	17509.063		
16	17505.409	36	17507.240	56	17509.173		
17	17505.533	37	17507.321	57	17509.304		
18	17505.594	38	17507.423	58	17509.371		
19	17505.675	39	17507.531	59	17509.428		
20	17505.746	40	17507.608	60	17509.490		

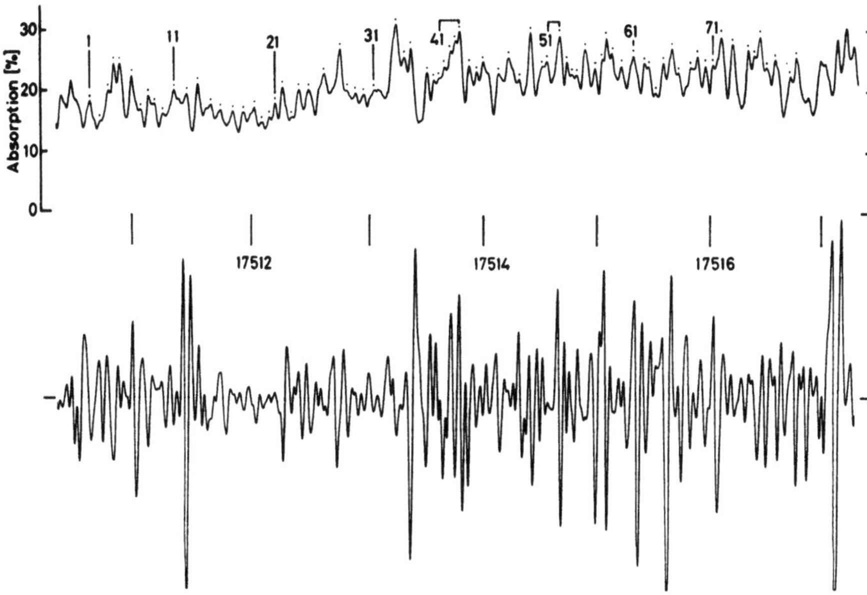

1	17510.647	21	17512.202	41	17513.793	61	17515.308
2	17510.735	22	17512.271	42	17513.872	62	17515.403
3	17510.808	23	17512.350	43	17513.942	63	17515.441
4	17510.858	24	17512.408	44	17514.000	64	17515.500
5	17510.907	25	17512.496	45	17514.125	65	17515.574
6	17511.004	26	17512.630	46	17514.221	66	17515.653
7	17511.073	27	17512.766	47	17514.312	67	17515.715
8	17511.141	28	17512.834	48	17514.415	68	17515.829
9	17511.189	29	17512.907	49	17514.509	69	17515.887
10	17511.283	30	17512.963	50	17514.552	70	17515.962
11	17511.356	31	17513.078	51	17514.666	71	17516.029
12	17511.457	32	17513.239	52	17514.731	72	17516.103
13	17511.554	33	17513.307	53	17514.775	73	17516.198
14	17511.656	34	17513.371	54	17514.814	74	17516.340
15	17511.736	35	17513.502	55	17514.891	75	17516.397
16	17511.844	36	17513.561	56	17514.972	76	17516.440
17	17511.934	37	17513.608	57	17515.034	77	17516.505
18	17512.029	38	17513.655	58	17515.078	78	17516.577
19	17512.091	39	17513.711	59	17515.121	79	17516.718
20	17512.157	40	17513.752	60	17515.201	80	17516.775

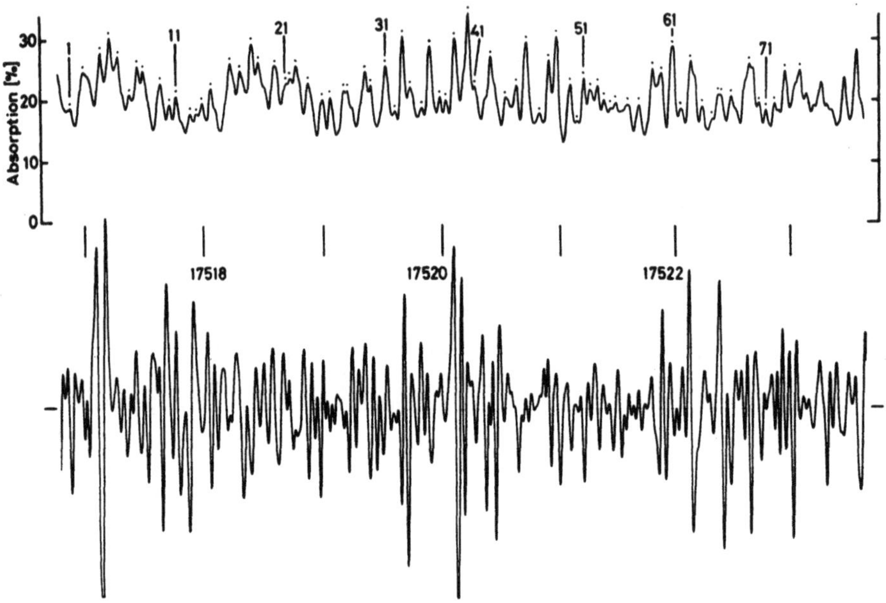

1	17516.886	21	17518.690	41	17520.280	61	17521.977
2	17516.996	22	17518.731	42	17520.417	62	17522.058
3	17517.138	23	17518.779	43	17520.545	63	17522.140
4	17517.219	24	17518.878	44	17520.634	64	17522.244
5	17517.290	25	17518.998	45	17520.714	65	17522.328
6	17517.395	26	17519.067	46	17520.831	66	17522.374
7	17517.463	27	17519.182	47	17520.912	67	17522.405
8	17517.511	28	17519.209	48	17520.973	68	17522.495
9	17517.655	29	17519.362	49	17521.098	69	17522.654
10	17517.727	30	17519.409	50	17521.157	70	17522.738
11	17517.785	31	17519.536	51	17521.215	71	17522.796
12	17517.895	32	17519.612	52	17521.269	72	17522.873
13	17517.949	33	17519.670	53	17521.332	73	17522.963
14	17517.999	34	17519.735	54	17521.393	74	17523.092
15	17518.071	35	17519.832	55	17521.444		
16	17518.238	36	17519.899	56	17521.487		
17	17518.326	37	17519.988	57	17521.594		
18	17518.417	38	17520.038	58	17521.687		
19	17518.480	39	17520.110	59	17521.803		
20	17518.600	40	17520.228	60	17521.877		

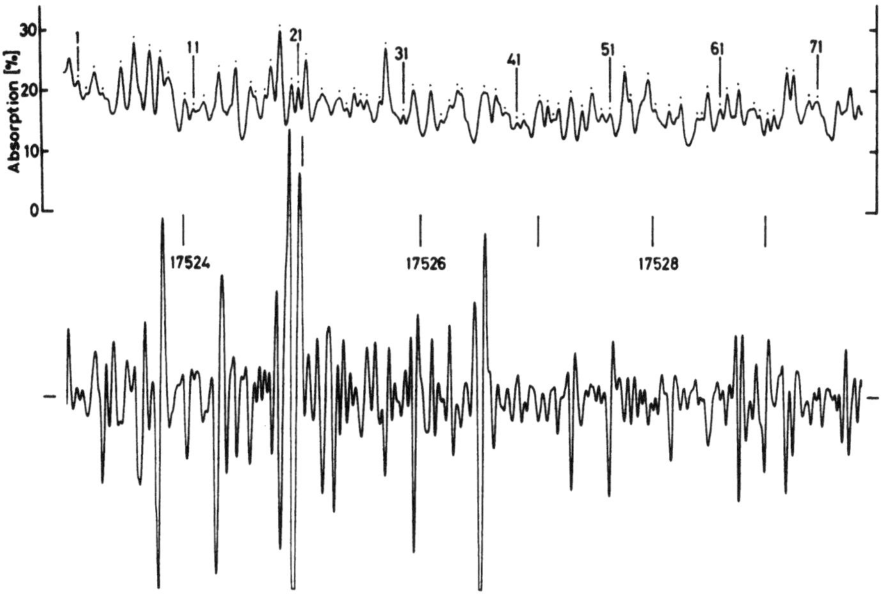

1	17523.158	21	17524.973	41	17526.832	61	17528.601
2	17523.225	22	17525.039	42	17526.887	62	17528.662
3	17523.287	23	17525.176	43	17527.014	63	17528.766
4	17523.347	24	17525.309	44	17527.088	64	17528.902
5	17523.499	25	17525.377	45	17527.134	65	17528.948
6	17523.607	26	17525.441	46	17527.180	66	17529.023
7	17523.735	27	17525.493	47	17527.287	67	17529.081
8	17523.820	28	17525.544	48	17527.390	68	17529.200
9	17523.892	29	17525.658	49	17527.474	69	17529.262
10	17524.022	30	17525.711	50	17527.559	70	17529.400
11	17524.093	31	17525.860	51	17527.644	71	17529.475
12	17524.180	32	17525.946	52	17527.765		
13	17524.301	33	17526.094	53	17527.808		
14	17524.440	34	17526.185	54	17527.960		
15	17524.565	35	17526.256	55	17528.023		
16	17524.607	36	17526.330	56	17528.143		
17	17524.690	37	17526.367	57	17528.251		
18	17524.742	38	17526.555	58	17528.394		
19	17524.818	39	17526.653	59	17528.438		
20	17524.916	40	17526.731	60	17528.495		

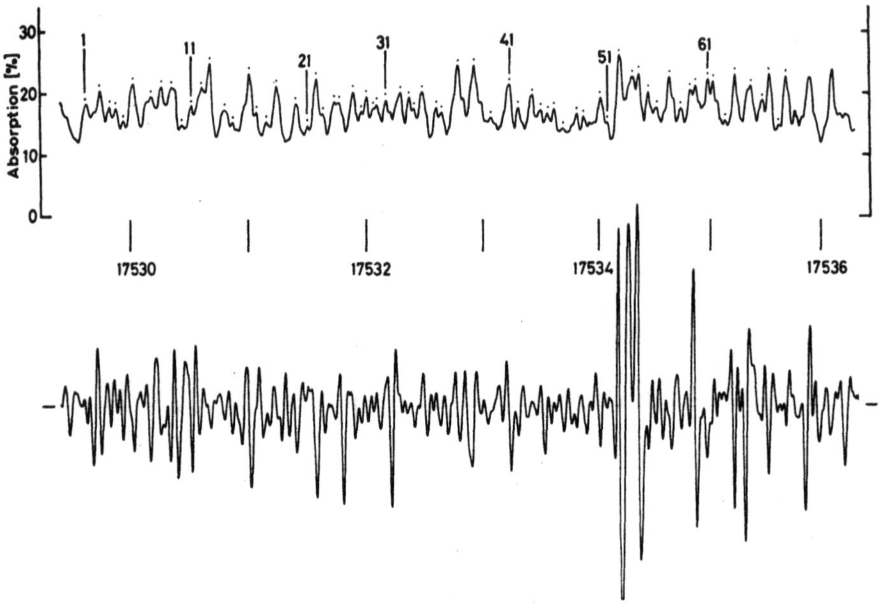

1	17529.653	21	17531.511	41	17533.220	61	17534.982
2	17529.758	22	17531.585	42	17533.298	62	17535.031
3	17529.831	23	17531.639	43	17533.416	63	17535.136
4	17529.883	24	17531.736	44	17533.485	64	17535.228
5	17529.949	25	17531.775	45	17533.553	65	17535.372
6	17530.029	26	17531.894	46	17533.611	66	17535.476
7	17530.179	27	17531.961	47	17533.684	67	17535.547
8	17530.265	28	17532.005	48	17533.814	68	17535.630
9	17530.369	29	17532.059	49	17533.869	69	17535.694
10	17530.439	30	17532.091	50	17534.018		
11	17530.515	31	17532.166	51	17534.078		
12	17530.611	32	17532.287	52	17534.182		
13	17530.671	33	17532.357	53	17534.297		
14	17530.810	34	17532.402	54	17534.355		
15	17530.874	35	17532.468	55	17534.442		
16	17531.010	36	17532.595	56	17534.521		
17	17531.073	37	17532.648	57	17534.632		
18	17531.159	38	17532.780	58	17534.732		
19	17531.240	39	17532.922	59	17534.823		
20	17531.408	40	17533.065	60	17534.869		

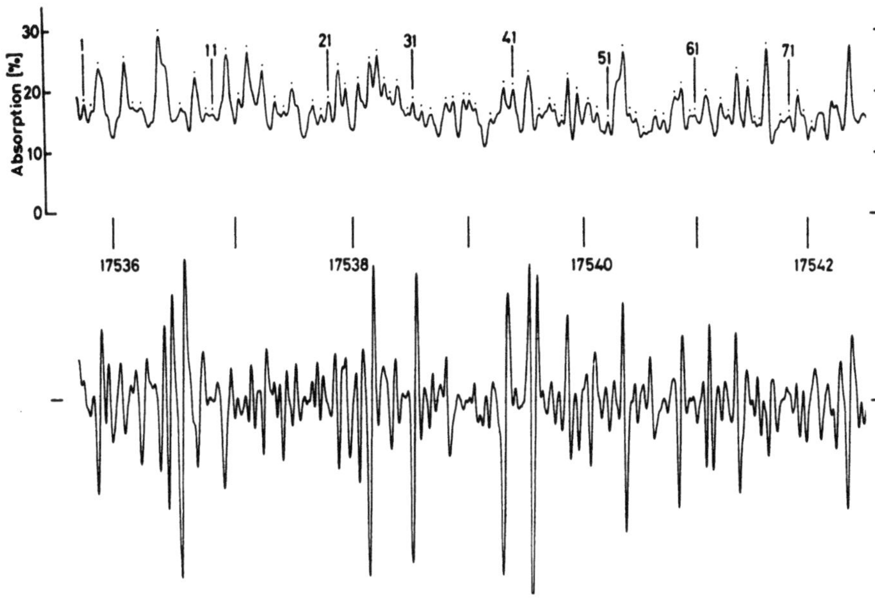

1	17535.783	21	17537.810	41	17539.388	61	17540.987
2	17535.842	22	17537.894	42	17539.519	62	17541.090
3	17535.901	23	17537.955	43	17539.614	63	17541.220
4	17536.110	24	17538.064	44	17539.702	64	17541.284
5	17536.174	25	17538.159	45	17539.747	65	17541.359
6	17536.233	26	17538.224	46	17539.802	66	17541.460
7	17536.393	27	17538.287	47	17539.859	67	17541.518
8	17536.579	28	17538.334	48	17539.946	68	17541.559
9	17536.698	29	17538.398	49	17540.034	69	17541.622
10	17536.789	30	17538.489	50	17540.118	70	17541.757
11	17536.839	31	17538.529	51	17540.219	71	17541.836
12	17536.954	32	17538.606	52	17540.354	72	17541.913
13	17537.054	33	17538.682	53	17540.408	73	17541.966
14	17537.128	34	17538.813	54	17540.464	74	17542.042
15	17537.251	35	17538.873	55	17540.542		
16	17537.361	36	17538.963	56	17540.636		
17	17537.436	37	17539.011	57	17540.711		
18	17537.504	38	17539.064	58	17540.818		
19	17537.678	39	17539.194	59	17540.871		
20	17537.747	40	17539.306	60	17540.946		

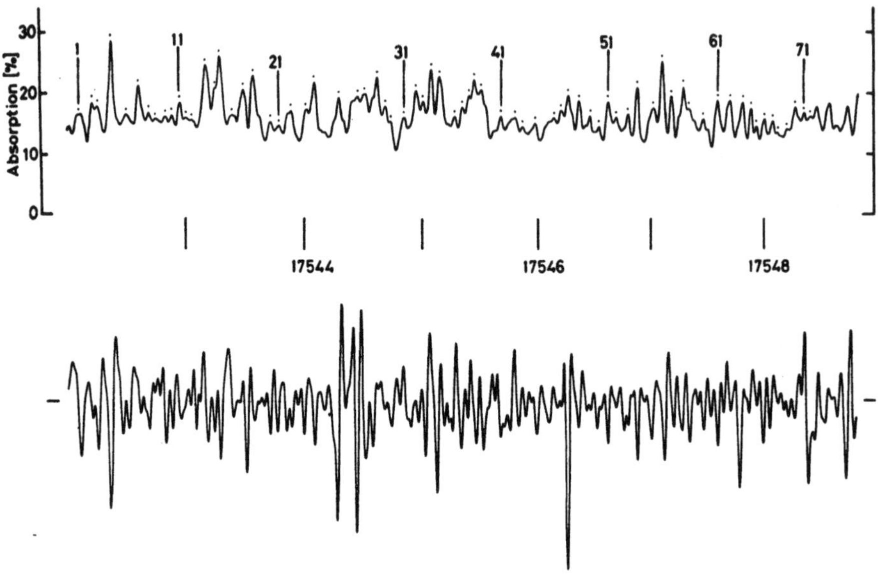

1	17542.121	21	17543.785	41	17545.685	61	17547.596
2	17542.224	22	17543.893	42	17545.807	62	17547.699
3	17542.269	23	17544.029	43	17545.871	63	17547.815
4	17542.386	24	17544.099	44	17545.981	64	17547.892
5	17542.514	25	17544.307	45	17546.139	65	17547.944
6	17542.617	26	17544.460	46	17546.204	66	17548.012
7	17542.698	27	17544.527	47	17546.266	67	17548.085
8	17542.759	28	17544.637	48	17546.363	68	17548.131
9	17542.834	29	17544.699	49	17546.457	69	17548.220
10	17542.895	30	17544.746	50	17546.539	70	17548.286
11	17542.960	31	17544.855	51	17546.627	71	17548.367
12	17543.012	32	17544.961	52	17546.693	72	17548.418
13	17543.065	33	17545.019	53	17546.802		
14	17543.178	34	17545.091	54	17546.888		
15	17543.257	35	17545.164	55	17547.030		
16	17543.298	36	17545.290	56	17547.108		
17	17543.418	37	17545.357	57	17547.190		
18	17543.499	38	17545.417	58	17547.301		
19	17543.575	39	17545.463	59	17547.343		
20	17543.718	40	17545.520	60	17547.470		

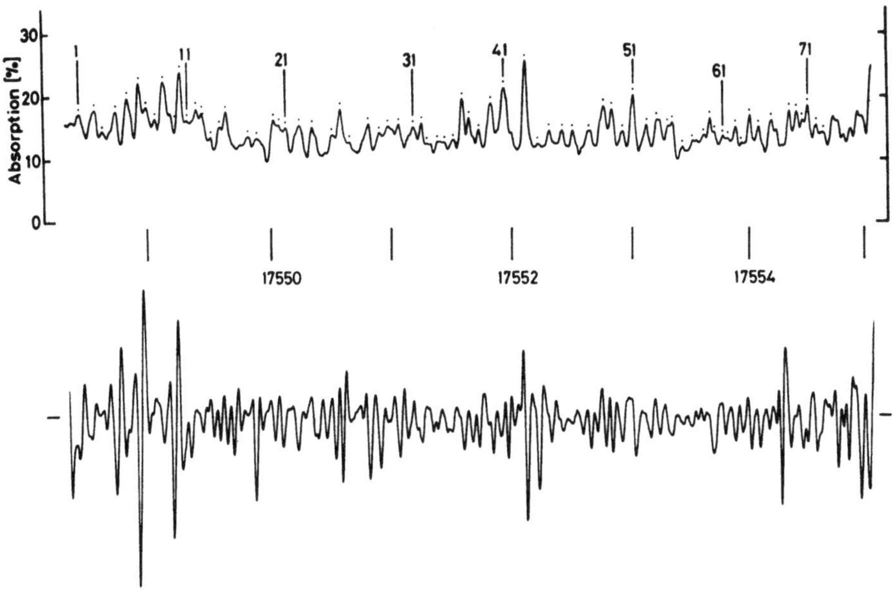

1	17548.474	21	17550.134	41	17551.941	61	17553.782
2	17548.593	22	17550.248	42	17552.115	62	17553.822
3	17548.651	23	17550.353	43	17552.221	63	17553.883
4	17548.757	24	17550.519	44	17552.311	64	17553.941
5	17548.848	25	17550.588	45	17552.416	65	17554.005
6	17548.939	26	17550.664	46	17552.506	66	17554.084
7	17549.001	27	17550.823	47	17552.645	67	17554.188
8	17549.069	28	17550.912	48	17552.766	68	17554.349
9	17549.136	29	17550.983	49	17552.834	69	17554.416
10	17549.270	30	17551.072	50	17552.917	70	17554.468
11	17549.324	31	17551.191	51	17553.010	71	17554.516
12	17549.401	32	17551.260	52	17553.124	72	17554.587
13	17549.444	33	17551.306	53	17553.212	73	17554.648
14	17549.516	34	17551.391	54	17553.302		
15	17549.590	35	17551.449	55	17553.335		
16	17549.640	36	17551.526	56	17553.424		
17	17549.818	37	17551.593	57	17553.519		
18	17549.894	38	17551.649	58	17553.606		
19	17550.035	39	17551.732	59	17553.661		
20	17550.079	40	17551.831	60	17553.704		

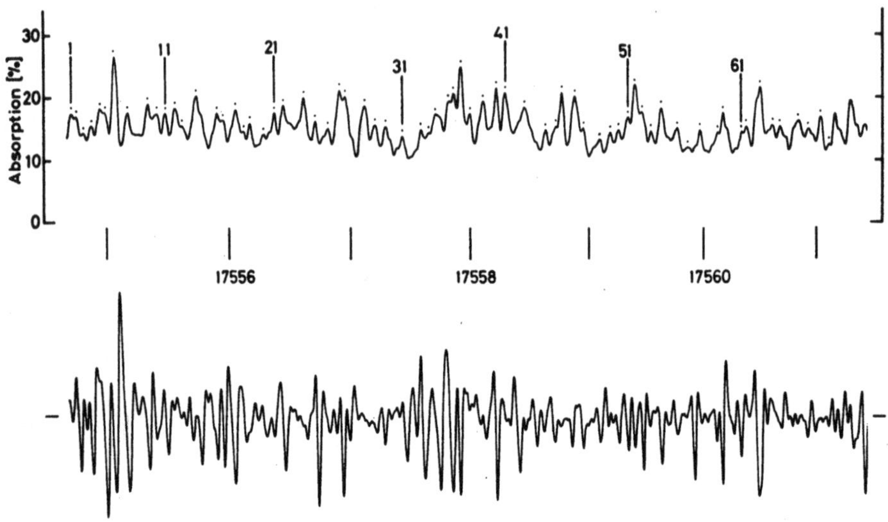

1	17554.729	21	17556.367	41	17558.297	61	17560.334
2	17554.767	22	17556.444	42	17558.459	62	17560.381
3	17554.824	23	17556.618	43	17558.643	63	17560.512
4	17554.878	24	17556.715	44	17558.723	64	17560.617
5	17554.947	25	17556.817	45	17558.773	65	17560.689
6	17554.990	26	17556.919	46	17558.882	66	17560.849
7	17555.066	27	17556.965	47	17558.956	67	17560.935
8	17555.175	28	17557.120	48	17559.096	68	17561.043
9	17555.340	29	17557.208	49	17559.190	69	17561.112
10	17555.408	30	17557.297	50	17559.259		
11	17555.480	31	17557.430	51	17559.343		
12	17555.560	32	17557.589	52	17559.403		
13	17555.621	33	17557.649	53	17559.468		
14	17555.727	34	17557.713	54	17559.547		
15	17555.898	35	17557.819	55	17559.635		
16	17555.951	36	17557.865	56	17559.772		
17	17556.059	37	17557.918	57	17559.874		
18	17556.124	38	17558.000	58	17559.975		
19	17556.171	39	17558.112	59	17560.124		
20	17556.281	40	17558.225	60	17560.174		

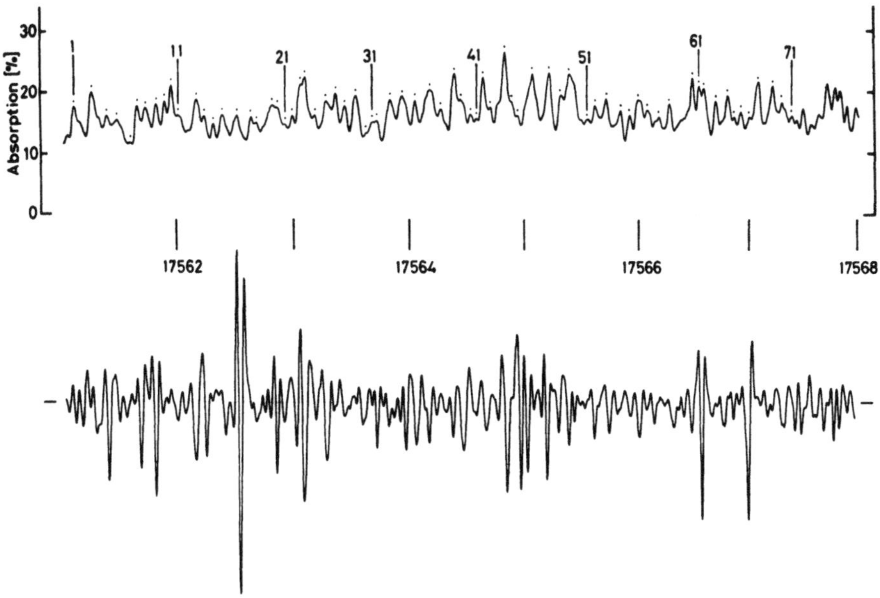

1	17561.161	21	17562.932	41	17564.586	61	17566.546
2	17561.301	22	17562.996	42	17564.643	62	17566.598
3	17561.425	23	17563.062	43	17564.703	63	17566.703
4	17561.510	24	17563.104	44	17564.834	64	17566.812
5	17561.621	25	17563.192	45	17564.895	65	17566.872
6	17561.674	26	17563.290	46	17564.941	66	17566.934
7	17561.744	27	17563.375	47	17565.071	67	17567.013
8	17561.834	28	17563.453	48	17565.212	68	17567.089
9	17561.906	29	17563.542	49	17565.323	69	17567.226
10	17561.966	30	17563.629	50	17565.392	70	17567.312
11	17562.020	31	17563.680	51	17565.547	71	17567.395
12	17562.171	32	17563.723	52	17565.624		
13	17562.238	33	17563.839	53	17565.722		
14	17562.316	34	17563.935	54	17565.845		
15	17562.389	35	17564.054	55	17565.925		
16	17562.511	36	17564.186	56	17566.003		
17	17562.626	37	17564.281	57	17566.081		
18	17562.680	38	17564.391	58	17566.186		
19	17562.814	39	17564.452	59	17566.274		
20	17562.856	40	17564.537	60	17566.489		

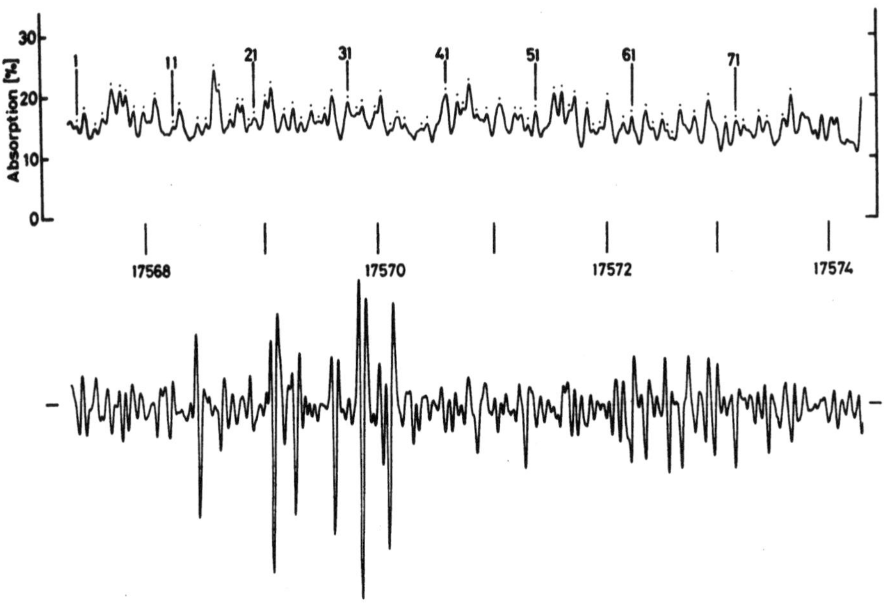

1	17567.450	21	17568.936	41	17570.596	61	17572.225
2	17567.510	22	17569.030	42	17570.689	62	17572.348
3	17567.588	23	17569.081	43	17570.738	63	17572.409
4	17567.653	24	17569.188	44	17570.790	64	17572.497
5	17567.733	25	17569.271	45	17570.863	65	17572.541
6	17567.809	26	17569.338	46	17570.941	66	17572.589
7	17567.856	27	17569.432	47	17571.038	67	17572.661
8	17567.914	28	17569.494	48	17571.179	68	17572.789
9	17567.997	29	17569.549	49	17571.237	69	17572.915
10	17568.095	30	17569.612	50	17571.296	70	17573.075
11	17568.243	31	17569.755	51	17571.364	71	17573.172
12	17568.299	32	17569.836	52	17571.529	72	17573.243
13	17568.447	33	17569.877	53	17571.599	73	17573.381
14	17568.517	34	17569.996	54	17571.666	74	17573.453
15	17568.591	35	17570.036	55	17571.709	75	17573.604
16	17568.630	36	17570.113	56	17571.823	76	17573.675
17	17568.732	37	17570.169	57	17571.879		
18	17568.795	38	17570.244	58	17571.936		
19	17568.837	39	17570.373	59	17572.006		
20	17568.893	40	17570.435	60	17572.146		

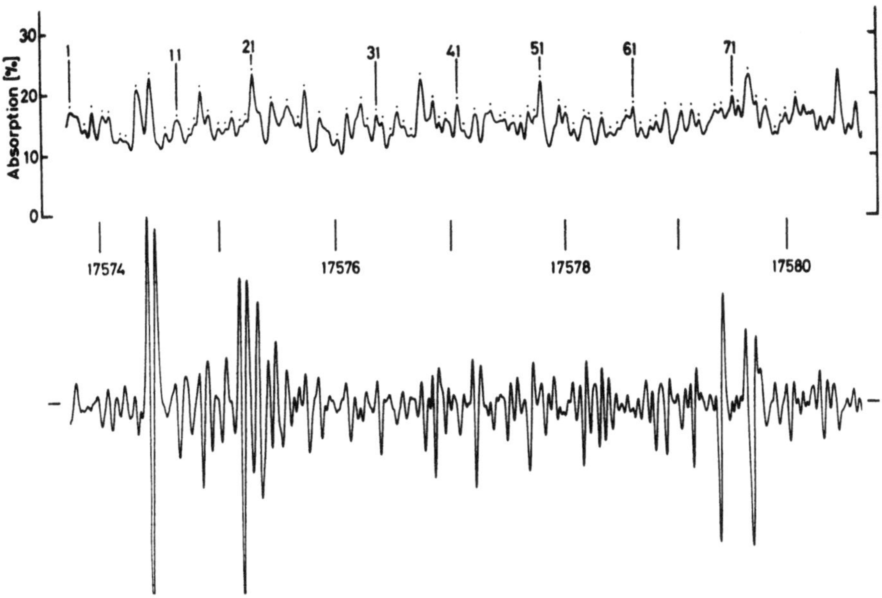

1	17573.768	21	17575.299	41	17577.074	61	17578.611
2	17573.888	22	17575.467	42	17577.128	62	17578.680
3	17573.950	23	17575.597	43	17577.228	63	17578.762
4	17574.038	24	17575.695	44	17577.354	64	17578.817
5	17574.089	25	17575.750	45	17577.438	65	17578.900
6	17574.189	26	17575.879	46	17577.482	66	17579.038
7	17574.235	27	17576.029	47	17577.557	67	17579.135
8	17574.326	28	17576.111	48	17577.615	68	17579.200
9	17574.435	29	17576.237	49	17577.686	69	17579.338
10	17574.575	30	17576.293	50	17577.743	70	17579.402
11	17574.665	31	17576.372	51	17577.791	71	17579.504
12	17574.760	32	17576.416	52	17577.957	72	17579.558
13	17574.814	33	17576.492	53	17578.014	73	17579.646
14	17574.864	34	17576.548	54	17578.087	74	17579.720
15	17574.931	35	17576.611	55	17578.181	75	17579.806
16	17575.012	36	17576.676	56	17578.225	76	17579.895
17	17575.074	37	17576.762	57	17578.333	77	17579.946
18	17575.125	38	17576.868	58	17578.402	78	17579.985
19	17575.197	39	17576.916	59	17578.483	79	17580.086
20	17575.243	40	17576.969	60	17578.556		

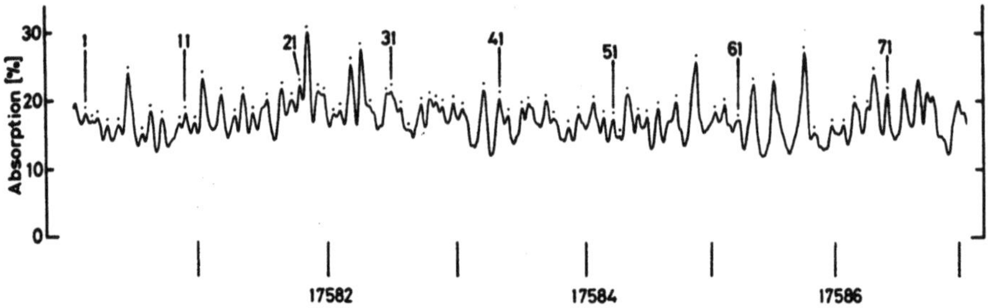

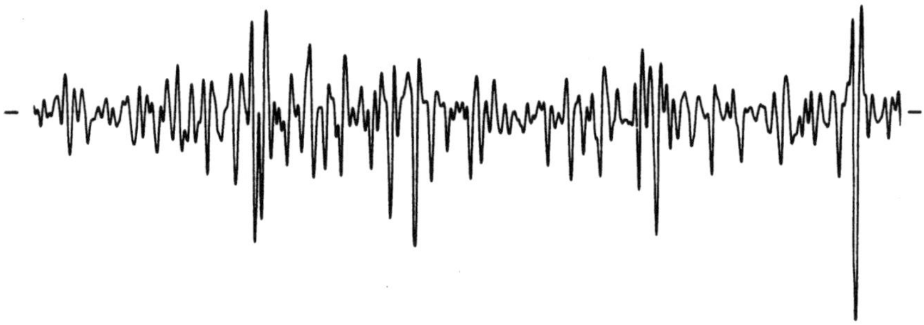

1	17580.164	21	17581.797	41	17583.325	61	17585.223
2	17580.209	22	17581.853	42	17583.394	62	17585.342
3	17580.249	23	17581.939	43	17583.505	63	17585.508
4	17580.323	24	17581.982	44	17583.554	64	17585.754
5	17580.405	25	17582.056	45	17583.689	65	17585.839
6	17580.478	26	17582.106	46	17583.752	66	17585.977
7	17580.592	27	17582.188	47	17583.867	67	17586.073
8	17580.652	28	17582.270	48	17583.947	68	17586.164
9	17580.738	29	17582.338	49	17584.061	69	17586.273
10	17580.869	30	17582.467	50	17584.140	70	17586.323
11	17580.915	31	17582.501	51	17584.218	71	17586.432
12	17580.986	32	17582.574	52	17584.266		
13	17581.045	33	17582.634	53	17584.332		
14	17581.188	34	17582.729	54	17584.416		
15	17581.293	35	17582.797	55	17584.487		
16	17581.361	36	17582.843	56	17584.575		
17	17581.438	37	17582.892	57	17584.715		
18	17581.550	38	17582.972	58	17584.876		
19	17581.659	39	17583.044	59	17585.036		
20	17581.733	40	17583.204	60	17585.111		

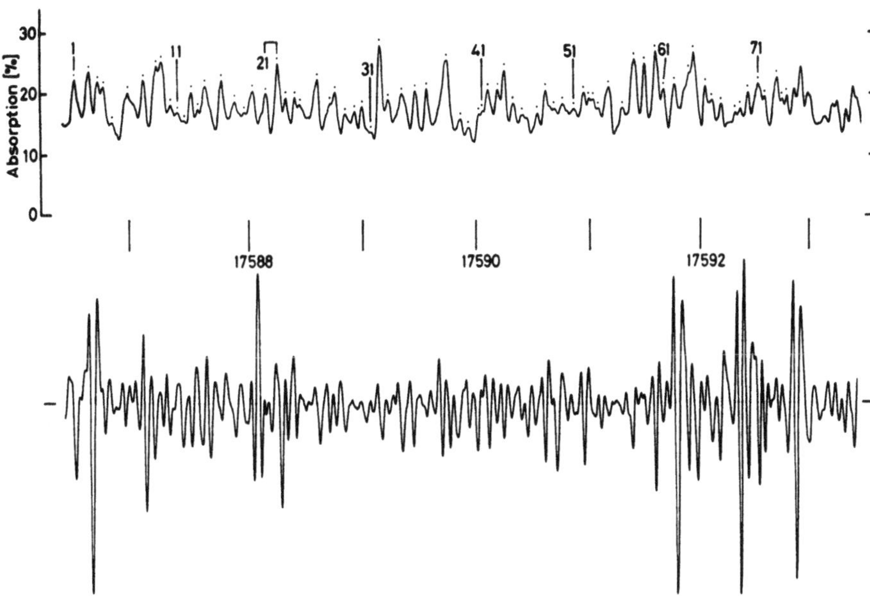

1	17586.557	21	17588.273	41	17590.065	61	17591.672
2	17586.678	22	17588.347	42	17590.114	62	17591.768
3	17586.752	23	17588.423	43	17590.199	63	17591.902
4	17586.795	24	17588.463	44	17590.255	64	17591.943
5	17586.871	25	17588.617	45	17590.330	65	17592.051
6	17586.998	26	17588.720	46	17590.408	66	17592.111
7	17587.134	27	17588.768	47	17590.551	67	17592.198
8	17587.240	28	17588.864	48	17590.615	68	17592.324
9	17587.291	29	17588.943	49	17590.693	69	17592.368
10	17587.369	30	17589.000	50	17590.759	70	17592.440
11	17587.419	31	17589.076	51	17590.866	71	17592.529
12	17587.492	32	17589.155	52	17590.949	72	17592.588
13	17587.547	33	17589.228	53	17591.000	73	17592.696
14	17587.597	34	17589.343	54	17591.040	74	17592.751
15	17587.657	35	17589.467	55	17591.092	75	17592.792
16	17587.796	36	17589.572	56	17591.178		
17	17587.907	37	17589.741	57	17591.297		
18	17587.984	38	17589.861	58	17591.405		
19	17588.052	39	17589.944	59	17591.497		
20	17588.172	40	17590.030	60	17591.601		

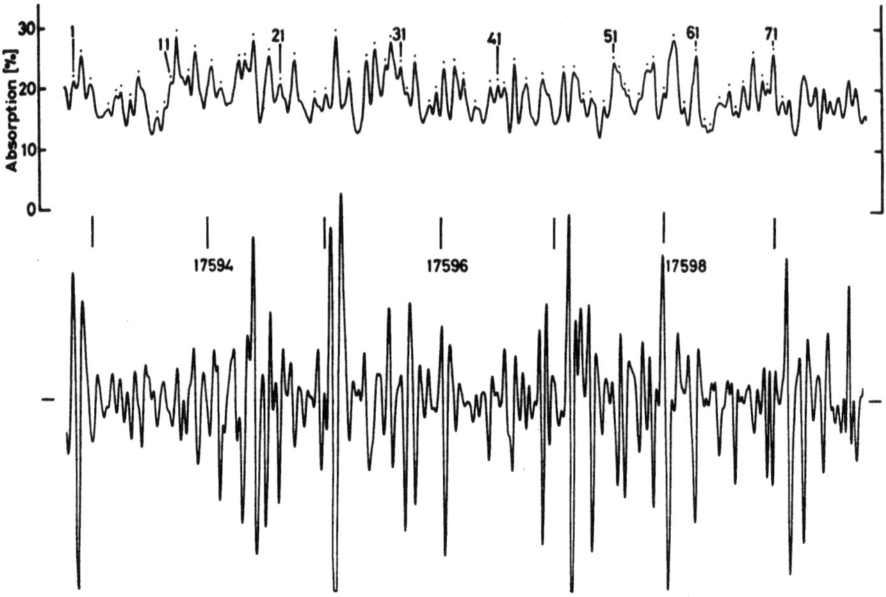

1	17592.860	21	17594.635	41	17596.506	61	17598.288
2	17592.920	22	17594.752	42	17596.561	62	17598.371
3	17592.997	23	17594.924	43	17596.645	63	17598.417
4	17593.155	24	17595.022	44	17596.752	64	17598.502
5	17593.226	25	17595.109	45	17596.899	65	17598.590
6	17593.266	26	17595.221	46	17597.084	66	17598.652
7	17593.348	27	17595.367	47	17597.179	67	17598.717
8	17593.417	28	17595.441	48	17597.264	68	17598.806
9	17593.577	29	17595.535	49	17597.341	69	17598.894
10	17593.640	30	17595.578	50	17597.457	70	17598.937
11	17593.697	31	17595.663	51	17597.544	71	17598.988
12	17593.743	32	17595.721	52	17597.591	72	17599.077
13	17593.842	33	17595.785	53	17597.638		
14	17593.903	34	17595.912	54	17597.674		
15	17594.044	35	17595.971	55	17597.775		
16	17594.128	36	17596.035	56	17597.849		
17	17594.283	37	17596.127	57	17597.902		
18	17594.336	38	17596.202	58	17597.999		
19	17594.407	39	17596.306	59	17598.086		
20	17594.537	40	17596.438	60	17598.184		

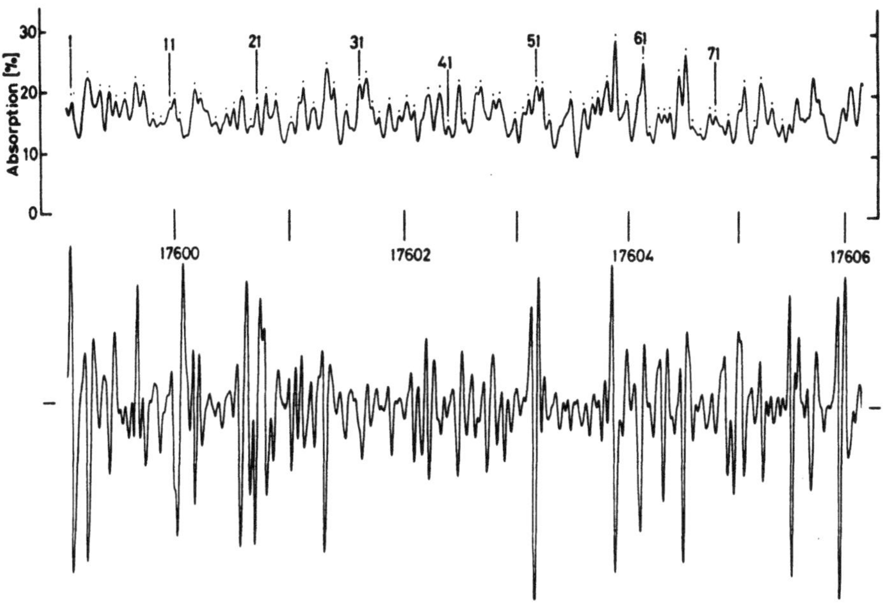

1	17599.140	21	17600.725	41	17602.394	61	17604.138
2	17599.275	22	17600.805	42	17602.486	62	17604.194
3	17599.377	23	17600.897	43	17602.540	63	17604.282
4	17599.452	24	17601.025	44	17602.636	64	17604.329
5	17599.506	25	17601.092	45	17602.679	65	17604.386
6	17599.585	26	17601.129	46	17602.793	66	17604.469
7	17599.682	27	17601.217	47	17602.844	67	17604.526
8	17599.743	28	17601.335	48	17602.986	68	17604.584
9	17599.830	29	17601.396	49	17603.069	69	17604.663
10	17599.896	30	17601.498	50	17603.113	70	17604.748
11	17599.973	31	17601.616	51	17603.183	71	17604.800
12	17600.011	32	17601.677	52	17603.231	72	17604.915
13	17600.057	33	17601.726	53	17603.294	73	17605.011
14	17600.187	34	17601.786	54	17603.487	74	17605.057
15	17600.235	35	17601.871	55	17603.604	75	17605.155
16	17600.366	36	17601.959	56	17603.680	76	17605.214
17	17600.461	37	17602.026	57	17603.734	77	17605.314
18	17600.523	38	17602.089	58	17603.816	78	17605.410
19	17600.595	39	17602.212	59	17603.884		
20	17600.666	40	17602.309	60	17603.986		

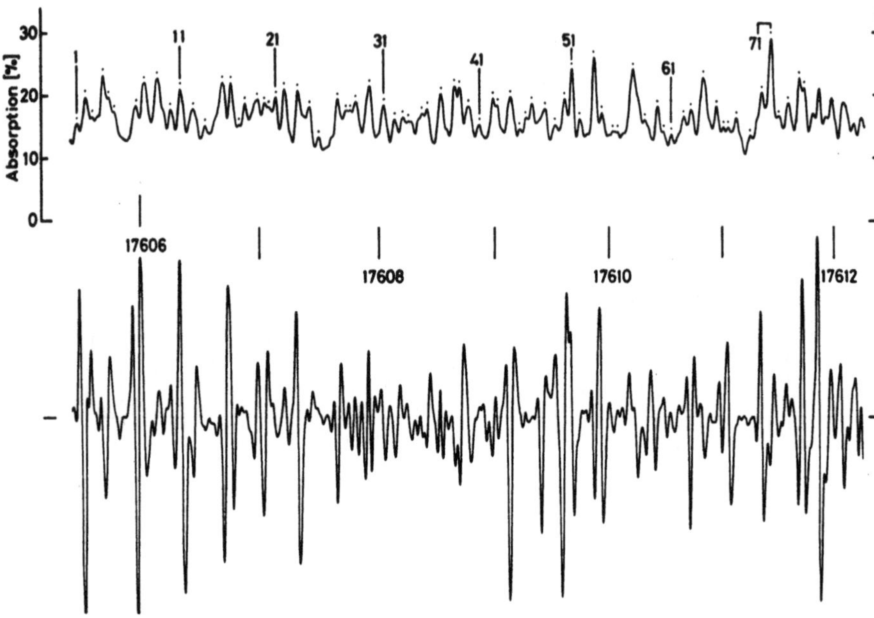

1	17605.481	21	17607.137	41	17608.870	61	17610.562
2	17605.551	22	17607.209	42	17608.989	62	17610.674
3	17605.606	23	17607.320	43	17609.026	63	17610.736
4	17605.700	24	17607.414	44	17609.141	64	17610.845
5	17605.747	25	17607.495	45	17609.225	65	17610.960
6	17605.793	26	17607.656	46	17609.281	66	17611.022
7	17605.976	27	17607.729	47	17609.331	67	17611.066
8	17606.047	28	17607.765	48	17609.445	68	17611.136
9	17606.157	29	17607.812	49	17609.535	69	17611.262
10	17606.265	30	17607.927	50	17609.615	70	17611.366
11	17606.343	31	17608.047	51	17609.679	71	17611.447
12	17606.446	32	17608.136	52	17609.749	72	17611.520
13	17606.545	33	17608.208	53	17609.877	73	17611.601
14	17606.695	34	17608.254	54	17609.947	74	17611.697
15	17606.766	35	17608.373	55	17610.044	75	17611.737
16	17606.837	36	17608.417	56	17610.086		
17	17606.884	37	17608.536	57	17610.222		
18	17606.988	38	17608.654	58	17610.326		
19	17607.045	39	17608.698	59	17610.440		
20	17607.082	40	17608.774	60	17610.491		

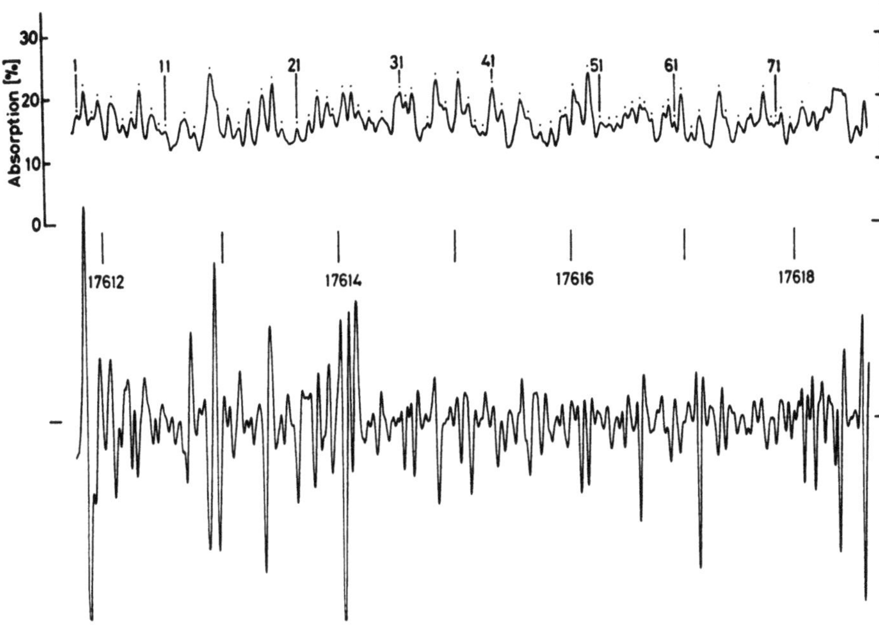

1	17611.810	21	17613.654	41	17615.333	61	17616.928
2	17611.872	22	17613.756	42	17615.417	62	17616.986
3	17611.932	23	17613.832	43	17615.573	63	17617.081
4	17611.984	24	17613.914	44	17615.648	64	17617.149
5	17612.098	25	17613.966	45	17615.752	65	17617.339
6	17612.191	26	17614.059	46	17615.846	66	17617.393
7	17612.260	27	17614.127	47	17615.928	67	17617.509
8	17612.327	28	17614.180	48	17615.964	68	17617.618
9	17612.421	29	17614.273	49	17616.037	69	17617.733
10	17612.489	30	17614.390	50	17616.167	70	17617.802
11	17612.540	31	17614.542	51	17616.279	71	17617.845
12	17612.701	32	17614.594	52	17616.350	72	17617.891
13	17612.776	33	17614.649	53	17616.429	73	17617.974
14	17612.916	34	17614.780	54	17616.497	74	17618.093
15	17613.067	35	17614.849	55	17616.552		
16	17613.156	36	17614.935	56	17616.624		
17	17613.242	37	17615.044	57	17616.670		
18	17613.356	38	17615.126	58	17616.724		
19	17613.443	39	17615.187	59	17616.829		
20	17613.532	40	17615.248	60	17616.878		

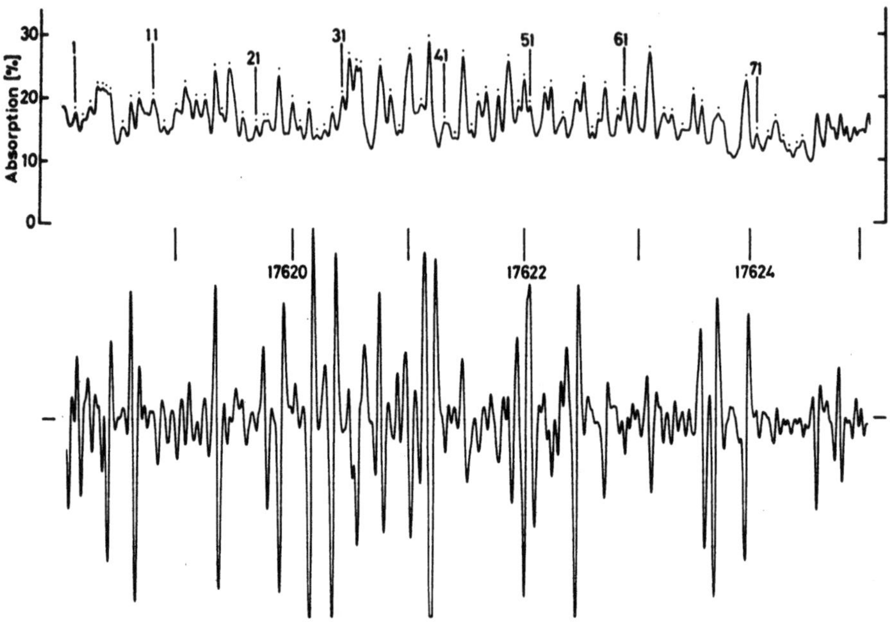

1	17618.187	21	17619.707	41	17621.331	61	17622.884
2	17618.247	22	17619.770	42	17621.422	62	17622.978
3	17618.301	23	17619.806	43	17621.488	63	17623.112
4	17618.367	24	17619.896	44	17621.558	64	17623.235
5	17618.406	25	17620.015	45	17621.620	65	17623.304
6	17618.446	26	17620.080	46	17621.687	66	17623.407
7	17618.480	27	17620.158	47	17621.790	67	17623.501
8	17618.581	28	17620.224	48	17621.876	68	17623.577
9	17618.653	29	17620.296	49	17621.963	69	17623.728
10	17618.718	30	17620.364	50	17622.009	70	17623.975
11	17618.836	31	17620.452	51	17622.064	71	17624.072
12	17618.923	32	17620.510	52	17622.192	72	17624.179
13	17619.025	33	17620.569	53	17622.245	73	17624.245
14	17619.104	34	17620.601	54	17622.353	74	17624.314
15	17619.197	35	17620.772	55	17622.461	75	17624.365
16	17619.270	36	17620.860	56	17622.531	76	17624.431
17	17619.358	37	17620.935	57	17622.602	77	17624.482
18	17619.412	38	17621.028	58	17622.658		
19	17619.481	39	17621.124	59	17622.720		
20	17619.594	40	17621.195	60	17622.828		

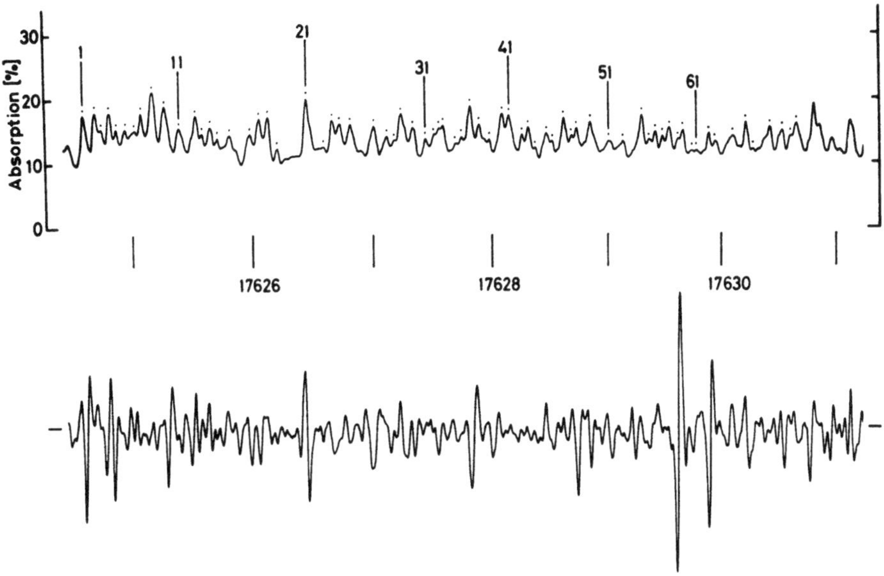

1	17624.610	21	17626.460	41	17628.154	61	17629.788
2	17624.711	22	17626.600	42	17628.259	62	17629.889
3	17624.757	23	17626.672	43	17628.309	63	17629.950
4	17624.827	24	17626.734	44	17628.373	64	17630.099
5	17624.874	25	17626.819	45	17628.474	65	17630.213
6	17624.953	26	17627.012	46	17628.515	66	17630.270
7	17625.029	27	17627.120	47	17628.620	67	17630.421
8	17625.081	28	17627.184	48	17628.688	68	17630.527
9	17625.178	29	17627.248	49	17628.731	69	17630.606
10	17625.278	30	17627.347	50	17628.854	70	17630.659
11	17625.398	31	17627.449	51	17629.020		
12	17625.535	32	17627.523	52	17629.139		
13	17625.580	33	17627.560	53	17629.300		
14	17625.653	34	17627.602	54	17629.366		
15	17625.714	35	17627.709	55	17629.413		
16	17625.813	36	17627.763	56	17629.481		
17	17625.981	37	17627.831	57	17629.539		
18	17626.063	38	17627.903	58	17629.612		
19	17626.138	39	17627.986	59	17629.661		
20	17626.214	40	17628.093	60	17629.729		

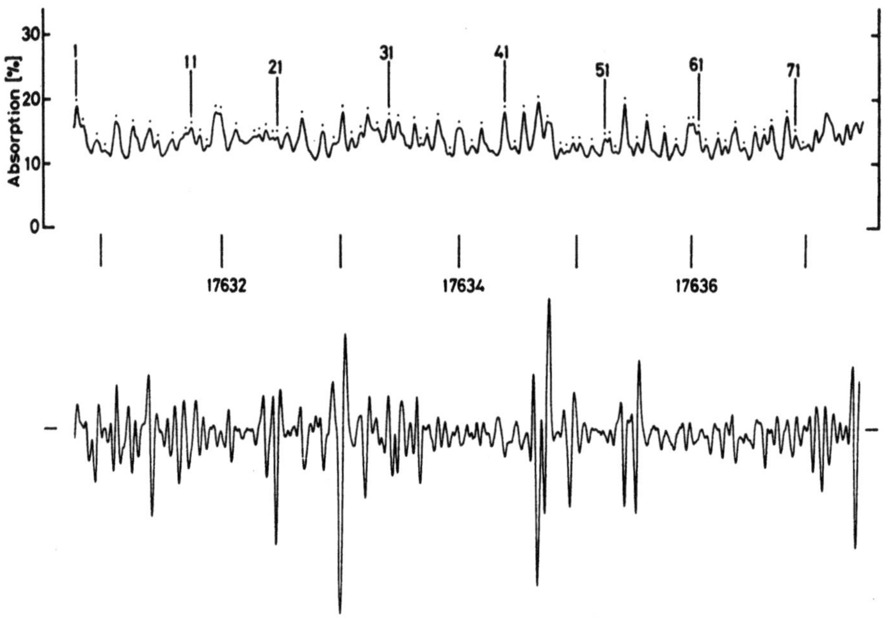

1	17630.824	21	17632.471	41	17634.383	61	17636.054
2	17630.868	22	17632.557	42	17634.471	62	17636.120
3	17630.970	23	17632.679	43	17634.546	63	17636.224
4	17631.041	24	17632.855	44	17634.669	64	17636.284
5	17631.134	25	17632.947	45	17634.745	65	17636.377
6	17631.279	26	17633.023	46	17634.862	66	17636.458
7	17631.416	27	17633.099	47	17634.913	67	17636.550
8	17631.480	28	17633.178	48	17634.967	68	17636.631
9	17631.600	29	17633.230	49	17635.022	69	17636.700
10	17631.709	30	17633.318	50	17635.131	70	17636.839
11	17631.755	31	17633.408	51	17635.242	71	17636.913
12	17631.820	32	17633.487	52	17635.279	72	17636.985
13	17631.891	33	17633.624	53	17635.333		
14	17631.962	34	17633.681	54	17635.406		
15	17631.999	35	17633.731	55	17635.520		
16	17632.130	36	17633.823	56	17635.597		
17	17632.273	37	17633.925	57	17635.752		
18	17632.319	38	17634.000	58	17635.857		
19	17632.371	39	17634.111	59	17635.969		
20	17632.424	40	17634.190	60	17636.008		

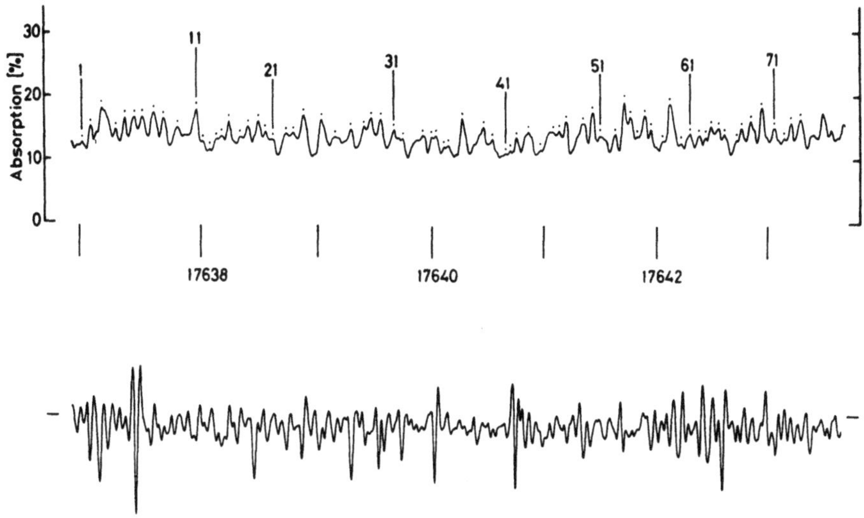

1	17637.028	21	17638.618	41	17640.641	61	17642.307
2	17637.095	22	17638.731	42	17640.687	62	17642.390
3	17637.189	23	17638.794	43	17640.746	63	17642.448
4	17637.306	24	17638.886	44	17640.858	64	17642.496
5	17637.376	25	17639.038	45	17640.957	65	17642.562
6	17637.455	26	17639.150	46	17641.070	66	17642.618
7	17637.518	27	17639.289	47	17641.136	67	17642.724
8	17637.621	28	17639.400	48	17641.193	68	17642.767
9	17637.703	29	17639.460	49	17641.344	69	17642.845
10	17637.819	30	17639.546	50	17641.431	70	17642.947
11	17637.973	31	17639.663	51	17641.494	71	17643.062
12	17638.025	32	17639.742	52	17641.627	72	17643.168
13	17638.079	33	17639.920	53	17641.712	73	17643.221
14	17638.133	34	17639.990	54	17641.764	74	17643.318
15	17638.184	35	17640.034	55	17641.829		
16	17638.248	36	17640.107	56	17641.895		
17	17638.340	37	17640.144	57	17641.947		
18	17638.410	38	17640.266	58	17642.051		
19	17638.497	39	17640.456	59	17642.119		
20	17638.560	40	17640.532	60	17642.229		

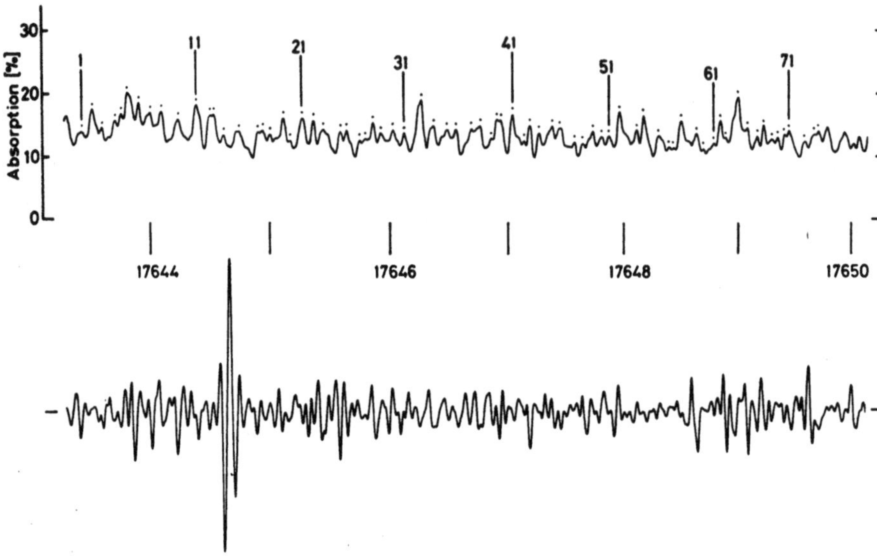

1	17643.429	21	17645.276	41	17647.043	61	17648.785
2	17643.525	22	17645.367	42	17647.140	62	17648.846
3	17643.605	23	17645.450	43	17647.189	63	17648.891
4	17643.719	24	17645.596	44	17647.270	64	17649.002
5	17643.773	25	17645.650	45	17647.384	65	17649.082
6	17643.826	26	17645.761	46	17647.448	66	17649.169
7	17643.918	27	17645.807	47	17647.577	67	17649.228
8	17644.006	28	17645.869	48	17647.647	68	17649.301
9	17644.104	29	17645.933	49	17647.736	69	17649.355
10	17644.239	30	17646.033	50	17647.810	70	17649.418
11	17644.392	31	17646.125	51	17647.872	71	17649.458
12	17644.510	32	17646.271	52	17647.965	72	17649.592
13	17644.543	33	17646.377	53	17648.115	73	17649.670
14	17644.633	34	17646.492	54	17648.176	74	17649.716
15	17644.754	35	17646.565	55	17648.309		
16	17644.913	36	17646.696	56	17648.404		
17	17644.950	37	17646.769	57	17648.439		
18	17645.010	38	17646.861	58	17648.508		
19	17645.119	39	17646.909	59	17648.639		
20	17645.177	40	17646.943	60	17648.699		

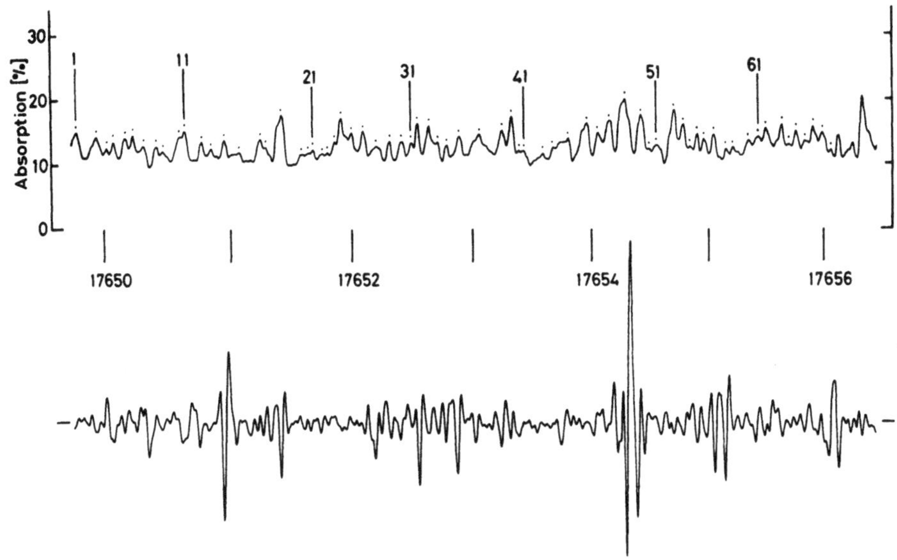

1	17649.798	21	17651.691	41	17653.432	61	17655.428
2	17649.950	22	17651.762	42	17653.592	62	17655.503
3	17650.032	23	17651.809	43	17653.679	63	17655.640
4	17650.081	24	17651.865	44	17653.806	64	17655.701
5	17650.175	25	17651.921	45	17653.961	65	17655.764
6	17650.235	26	17652.007	46	17654.062	66	17655.843
7	17650.316	27	17652.103	47	17654.153	67	17655.919
8	17650.417	28	17652.215	48	17654.202	68	17655.996
9	17650.476	29	17652.319	49	17654.429	69	17656.068
10	17650.605	30	17652.426	50	17654.494		
11	17650.646	31	17652.502	51	17654.554		
12	17650.780	32	17652.556	52	17654.709		
13	17650.855	33	17652.650	53	17654.790		
14	17650.959	34	17652.722	54	17654.852		
15	17651.078	35	17652.796	55	17654.911		
16	17651.252	36	17652.900	56	17654.969		
17	17651.306	37	17653.065	57	17655.053		
18	17651.433	38	17653.250	58	17655.166		
19	17651.595	39	17653.332	59	17655.223		
20	17651.650	40	17653.387	60	17655.352		

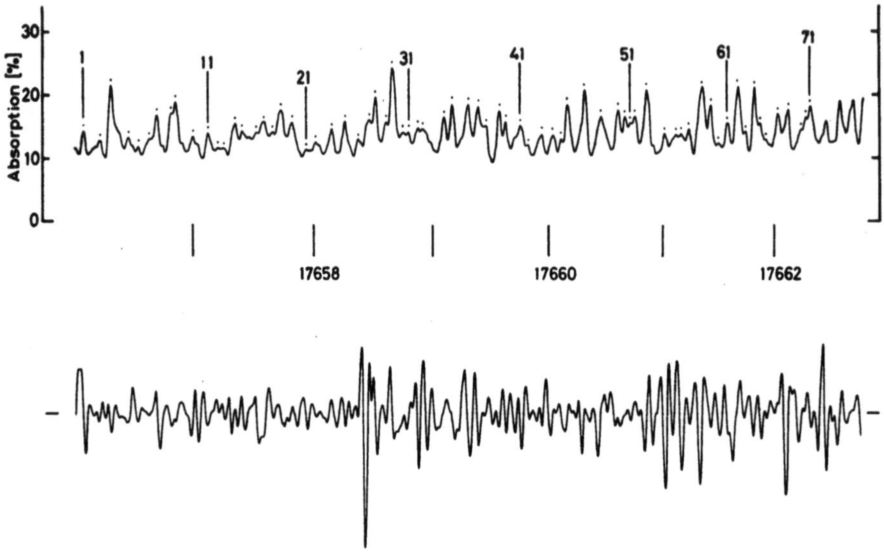

1	17656.131	21	17657.943	41	17659.754	61	17661.570
2	17656.255	22	17658.024	42	17659.823	62	17661.669
3	17656.341	23	17658.159	43	17659.939	63	17661.735
4	17656.486	24	17658.271	44	17660.038	64	17661.821
5	17656.571	25	17658.381	45	17660.109	65	17661.868
6	17656.666	26	17658.475	46	17660.161	66	17661.920
7	17656.720	27	17658.527	47	17660.313	67	17662.027
8	17656.834	28	17658.616	48	17660.456	68	17662.116
9	17656.871	29	17658.666	49	17660.603	69	17662.231
10	17657.007	30	17658.756	50	17660.662	70	17662.279
11	17657.130	31	17658.802	51	17660.711	71	17662.318
12	17657.215	32	17658.882	52	17660.750		
13	17657.261	33	17658.922	53	17660.854		
14	17657.359	34	17659.103	54	17661.010		
15	17657.414	35	17659.171	55	17661.115		
16	17657.534	36	17659.309	56	17661.163		
17	17657.593	37	17659.395	57	17661.231		
18	17657.666	38	17659.463	58	17661.348		
19	17657.732	39	17659.575	59	17661.419		
20	17657.825	40	17659.628	60	17661.488		

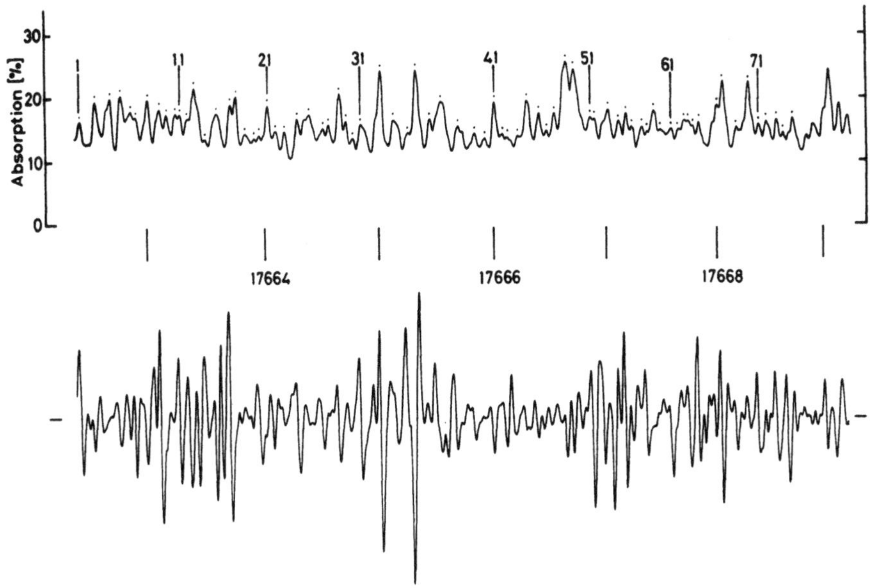

1	17662.457	21	17664.036	41	17666.010	61	17667.587
2	17662.583	22	17664.104	42	17666.078	62	17667.662
3	17662.705	23	17664.185	43	17666.123	63	17667.705
4	17662.796	24	17664.295	44	17666.219	64	17667.743
5	17662.875	25	17664.393	45	17666.292	65	17667.783
6	17662.916	26	17664.516	46	17666.398	66	17667.842
7	17663.019	27	17664.562	47	17666.466	67	17668.011
8	17663.114	28	17664.658	48	17666.535	68	17668.063
9	17663.171	29	17664.715	49	17666.642	69	17668.188
10	17663.253	30	17664.773	50	17666.712	70	17668.295
11	17663.290	31	17664.841	51	17666.863	71	17668.393
12	17663.345	32	17665.018	52	17666.903	72	17668.466
13	17663.408	33	17665.126	53	17667.024	73	17668.560
14	17663.500	34	17665.256	54	17667.117	74	17668.626
15	17663.598	35	17665.325	55	17667.180	75	17668.715
16	17663.711	36	17665.443	56	17667.235		
17	17663.757	37	17665.553	57	17667.324		
18	17663.840	38	17665.696	58	17667.368		
19	17663.921	39	17665.836	59	17667.431		
20	17663.959	40	17665.924	60	17667.494		

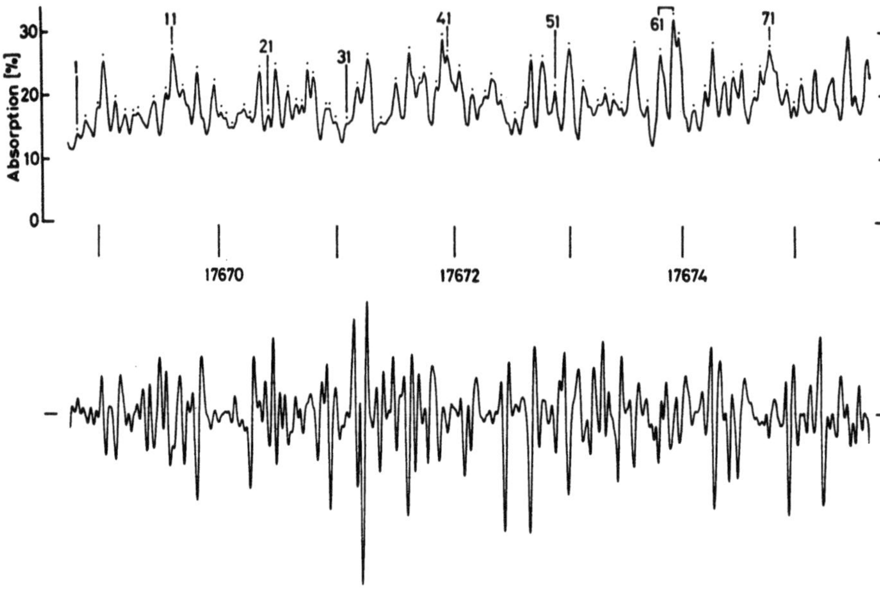

1	17668.845	21	17670.431	41	17671.945	61	17673.921
2	17668.907	22	17670.492	42	17672.046	62	17673.971
3	17668.996	23	17670.591	43	17672.157	63	17674.096
4	17669.052	24	17670.664	44	17672.264	64	17674.199
5	17669.147	25	17670.715	45	17672.331	65	17674.271
6	17669.228	26	17670.765	46	17672.407	66	17674.373
7	17669.304	27	17670.815	47	17672.527	67	17674.447
8	17669.339	28	17670.927	48	17672.613	68	17674.525
9	17669.467	29	17670.956	49	17672.666	69	17674.642
10	17669.572	30	17671.005	50	17672.765	70	17674.697
11	17669.630	31	17671.091	51	17672.878	71	17674.780
12	17669.709	32	17671.184	52	17672.998	72	17674.937
13	17669.826	33	17671.275	53	17673.119	73	17675.000
14	17669.972	34	17671.395	54	17673.251	74	17675.064
15	17670.029	35	17671.504	55	17673.313		
16	17670.072	36	17671.615	56	17673.389		
17	17670.126	37	17671.703	57	17673.456		
18	17670.228	38	17671.742	58	17673.577		
19	17670.276	39	17671.842	59	17673.682		
20	17670.356	40	17671.897	60	17673.804		

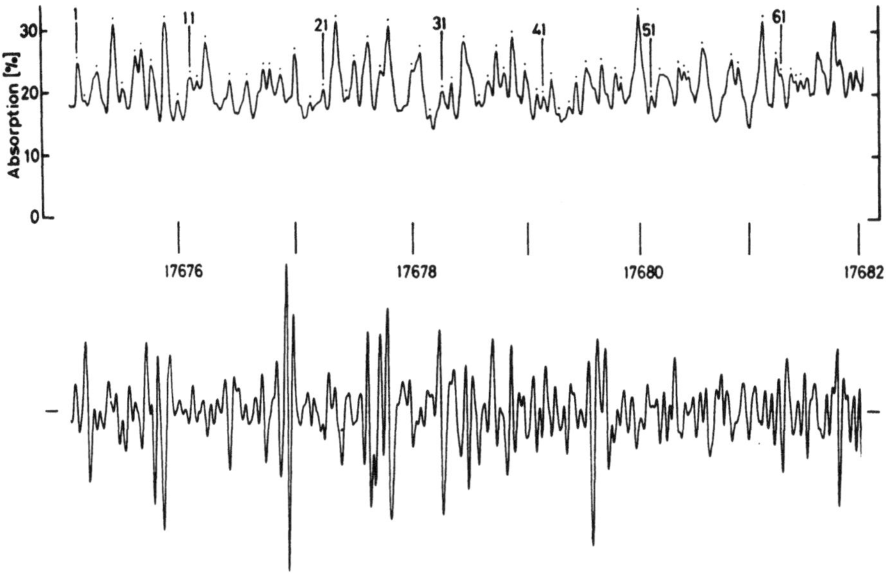

1	17675.191	21	17677.245	41	17679.150	61	17681.289
2	17675.244	22	17677.361	42	17679.220	62	17681.378
3	17675.346	23	17677.446	43	17679.277	63	17681.424
4	17675.482	24	17677.516	44	17679.380	64	17681.477
5	17675.548	25	17677.630	45	17679.435		
6	17675.653	26	17677.742	46	17679.530		
7	17675.710	27	17677.807	47	17679.667		
8	17675.786	28	17678.006	48	17679.790		
9	17675.903	29	17678.075	49	17679.838		
10	17676.003	30	17678.161	50	17679.989		
11	17676.111	31	17678.265	51	17680.107		
12	17676.164	32	17678.349	52	17680.187		
13	17676.244	33	17678.452	53	17680.351		
14	17676.448	34	17678.592	54	17680.399		
15	17676.592	35	17678.667	55	17680.440		
16	17676.732	36	17678.738	56	17680.580		
17	17676.786	37	17678.803	57	17680.841		
18	17676.874	38	17678.876	58	17680.897		
19	17677.003	39	17678.984	59	17681.120		
20	17677.127	40	17679.089	60	17681.242		

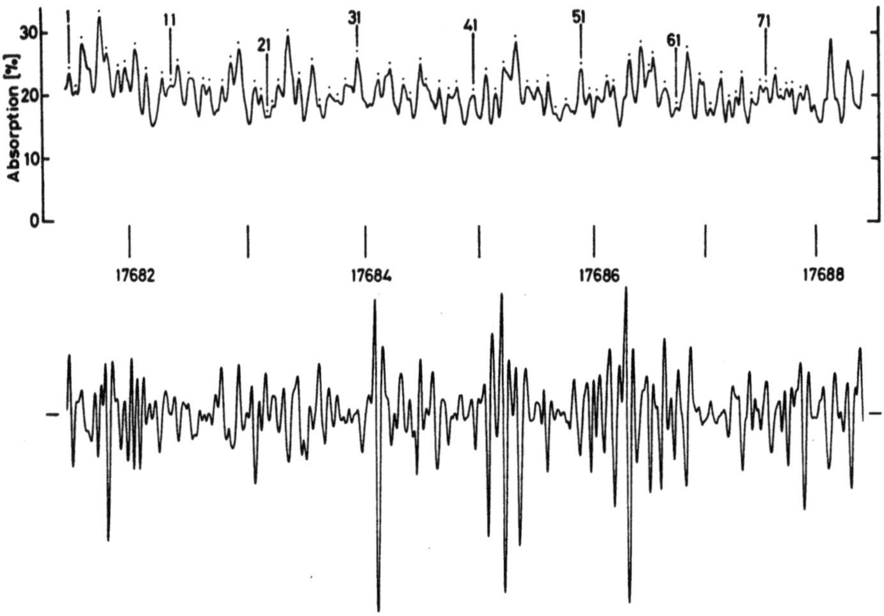

1	17681.534	21	17683.177	41	17684.956	61	17686.746
2	17681.584	22	17683.219	42	17685.066	62	17686.848
3	17681.630	23	17683.265	43	17685.148	63	17686.957
4	17681.775	24	17683.347	44	17685.221	64	17687.045
5	17681.831	25	17683.443	45	17685.332	65	17687.145
6	17681.925	26	17683.558	46	17685.430	66	17687.214
7	17681.979	27	17683.622	47	17685.520	67	17687.279
8	17682.060	28	17683.707	48	17685.607	68	17687.335
9	17682.150	29	17683.778	49	17685.682	69	17687.424
10	17682.285	30	17683.839	50	17685.760	70	17687.503
11	17682.360	31	17683.942	51	17685.896	71	17687.552
12	17682.420	32	17684.125	52	17685.967	72	17687.639
13	17682.512	33	17684.232	53	17686.036	73	17687.695
14	17682.627	34	17684.348	54	17686.127	74	17687.739
15	17682.681	35	17684.402	55	17686.181	75	17687.791
16	17682.789	36	17684.499	56	17686.330	76	17687.874
17	17682.866	37	17684.544	57	17686.429		
18	17682.940	38	17684.664	58	17686.502		
19	17683.077	39	17684.747	59	17686.543		
20	17683.128	40	17684.822	60	17686.657		

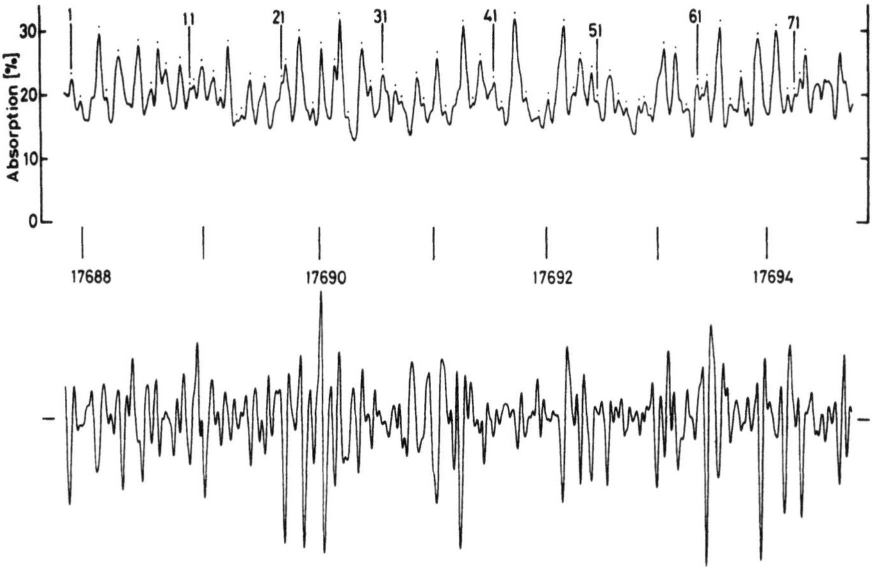

1	17687.931	21	17689.685	41	17691.537	61	17693.359
2	17687.997	22	17689.720	42	17691.603	62	17693.406
3	17688.149	23	17689.837	43	17691.720	63	17693.452
4	17688.206	24	17689.954	44	17691.819	64	17693.570
5	17688.304	25	17690.028	45	17691.933	65	17693.665
6	17688.466	26	17690.145	46	17692.020	66	17693.763
7	17688.577	27	17690.196	47	17692.155	67	17693.831
8	17688.631	28	17690.255	48	17692.244	68	17693.915
9	17688.697	29	17690.382	49	17692.299	69	17694.083
10	17688.820	30	17690.457	50	17692.401	70	17694.199
11	17688.898	31	17690.564	51	17692.455	71	17694.265
12	17688.933	32	17690.671	52	17692.571	72	17694.316
13	17689.003	33	17690.722	53	17692.647	73	17694.368
14	17689.095	34	17690.862	54	17692.720		
15	17689.152	35	17690.921	55	17692.832		
16	17689.217	36	17691.040	56	17692.884		
17	17689.295	37	17691.105	57	17692.936		
18	17689.336	38	17691.270	58	17693.061		
19	17689.402	39	17691.348	59	17693.162		
20	17689.530	40	17691.414	60	17693.267		

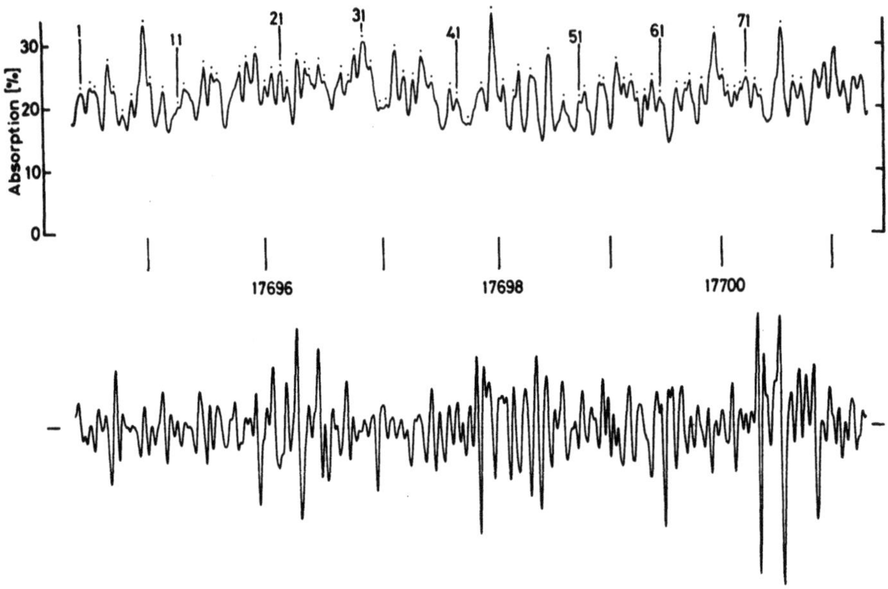

1	17694.471	21	17696.149	41	17697.641	61	17699.431
2	17694.547	22	17696.205	42	17697.741	62	17699.581
3	17694.584	23	17696.287	43	17697.858	63	17699.662
4	17694.694	24	17696.357	44	17697.948	64	17699.700
5	17694.744	25	17696.389	45	17698.044	65	17699.810
6	17694.815	26	17696.472	46	17698.139	66	17699.926
7	17694.894	27	17696.520	47	17698.178	67	17699.995
8	17694.983	28	17696.659	48	17698.284	68	17700.052
9	17695.040	29	17696.721	49	17698.439	69	17700.116
10	17695.145	30	17696.773	50	17698.569	70	17700.166
11	17695.281	31	17696.840	51	17698.708	71	17700.217
12	17695.339	32	17696.906	52	17698.757	72	17700.299
13	17695.489	33	17696.990	53	17698.886	73	17700.354
14	17695.556	34	17697.044	54	17698.923	74	17700.542
15	17695.601	35	17697.112	55	17699.040	75	17700.656
16	17695.800	36	17697.188	56	17699.105	76	17700.732
17	17695.861	37	17697.265	57	17699.172		
18	17695.941	38	17697.334	58	17699.227		
19	17696.013	39	17697.423	59	17699.297		
20	17696.076	40	17697.581	60	17699.358		

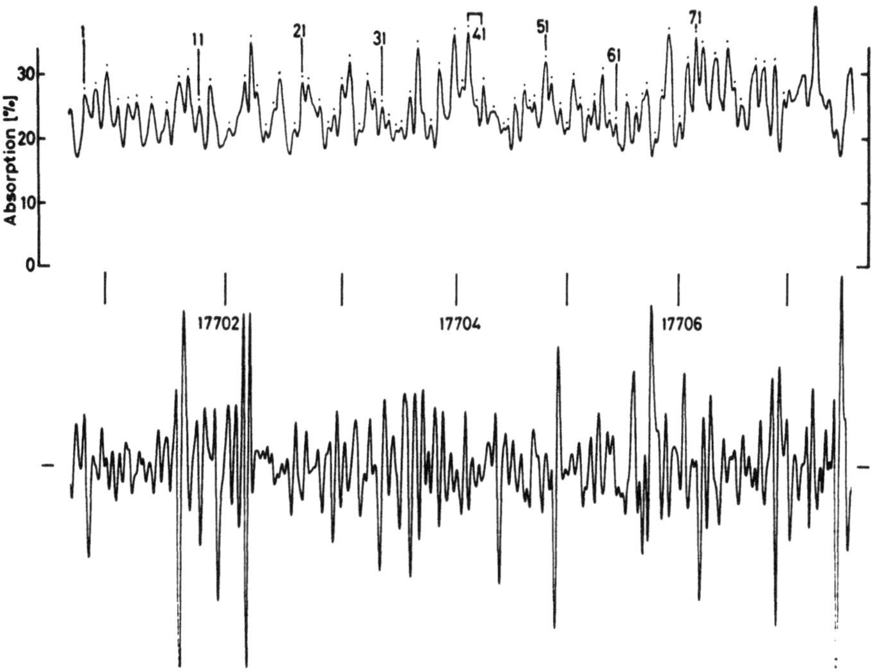

1	17700.847	21	17702.663	41	17704.115	61	17705.439
2	17700.931	22	17702.716	42	17704.189	62	17705.540
3	17701.028	23	17702.814	43	17704.251	63	17705.620
4	17701.118	24	17702.889	44	17704.342	64	17705.680
5	17701.203	25	17702.942	45	17704.425	65	17705.724
6	17701.280	26	17703.012	46	17704.462	66	17705.794
7	17701.397	27	17703.076	47	17704.525	67	17705.852
8	17701.530	28	17703.155	48	17704.611	68	17705.912
9	17701.632	29	17703.234	49	17704.659	69	17706.017
10	17701.706	30	17703.296	50	17704.701	70	17706.091
11	17701.793	31	17703.360	51	17704.799	71	17706.174
12	17701.885	32	17703.417	52	17704.856	72	17706.235
13	17702.040	33	17703.484	53	17704.934	73	17706.350
14	17702.172	34	17703.523	54	17704.998	74	17706.457
15	17702.227	35	17703.609	55	17705.050	75	17706.514
16	17702.275	36	17703.677	56	17705.116	76	17706.573
17	17702.351	37	17703.789	57	17705.191	77	17706.702
18	17702.426	38	17703.867	58	17705.244	78	17706.784
19	17702.472	39	17703.991	59	17705.316	79	17706.885
20	17702.595	40	17704.064	60	17705.376	80	17706.973

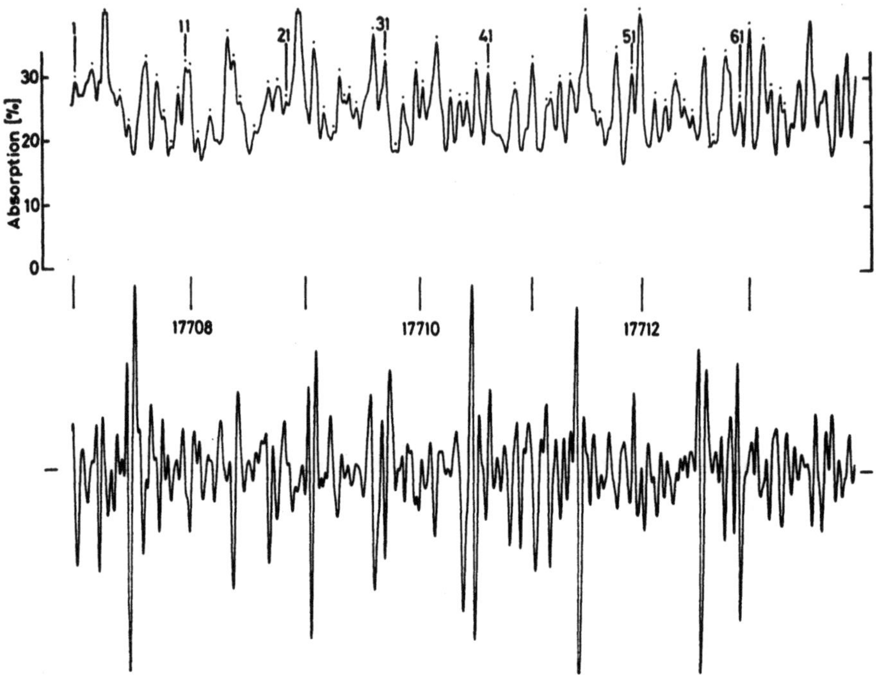

1	17707.030	21	17708.838	41	17710.600	61	17712.913
2	17707.158	22	17708.942	42	17710.838	62	17712.984
3	17707.262	23	17709.076	43	17710.998	63	17713.117
4	17707.401	24	17709.164	44	17711.130	64	17713.196
5	17707.476	25	17709.249	45	17711.162	65	17713.284
6	17707.617	26	17709.302	46	17711.250	66	17713.320
7	17707.712	27	17709.345	47	17711.347		
8	17707.777	28	17709.383	48	17711.476		
9	17707.839	29	17709.449	49	17711.621		
10	17707.895	30	17709.590	50	17711.762		
11	17707.963	31	17709.693	51	17711.902		
12	17707.994	32	17709.782	52	17711.975		
13	17708.071	33	17709.849	53	17712.122		
14	17708.182	34	17709.963	54	17712.217		
15	17708.338	35	17710.022	55	17712.314		
16	17708.390	36	17710.142	56	17712.396		
17	17708.446	37	17710.265	57	17712.473		
18	17708.569	38	17710.347	58	17712.578		
19	17708.685	39	17710.410	59	17712.666		
20	17708.767	40	17710.495	60	17712.770		

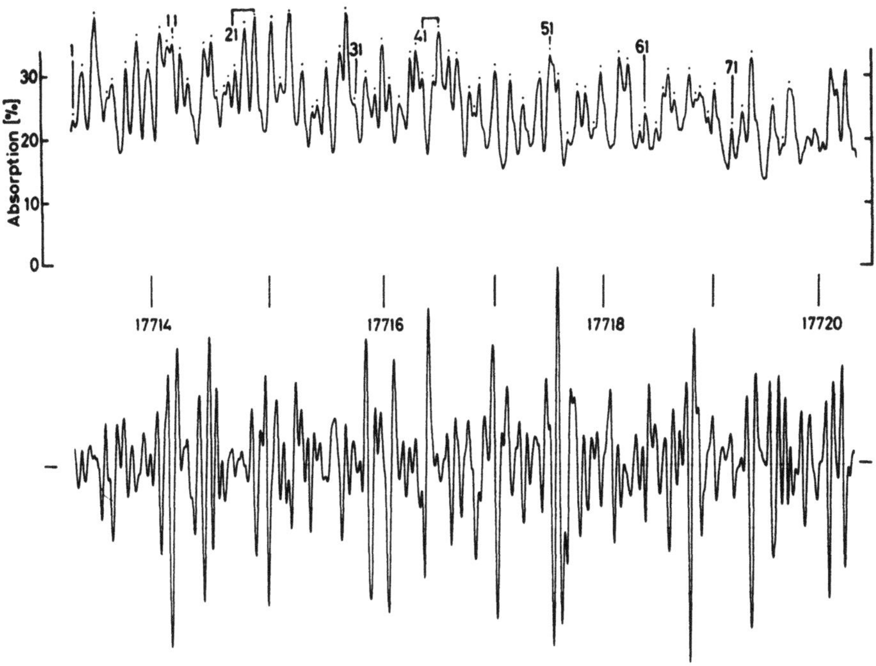

1	17713.385	21	17714.887	41	17716.512	61	17718.384
2	17713.451	22	17715.035	42	17716.601	62	17718.481
3	17713.538	23	17715.119	43	17716.675	63	17718.547
4	17713.646	24	17715.195	44	17716.790	64	17718.588
5	17713.684	25	17715.308	45	17716.828	65	17718.646
6	17713.803	26	17715.396	46	17716.875	66	17718.787
7	17713.893	27	17715.441	47	17717.013	67	17718.844
8	17713.988	28	17715.519	48	17717.154	68	17718.884
9	17714.085	29	17715.637	49	17717.268	69	17718.962
10	17714.149	30	17715.689	50	17717.419	70	17719.020
11	17714.192	31	17715.763	51	17717.519	71	17719.183
12	17714.260	32	17715.859	52	17717.598	72	17719.277
13	17714.323	33	17715.941	53	17717.677	73	17719.367
14	17714.458	34	17716.007	54	17717.766	74	17719.555
15	17714.517	35	17716.072	55	17717.839	75	17719.655
16	17714.563	36	17716.152	56	17717.921	76	17719.709
17	17714.626	37	17716.254	57	17717.981		
18	17714.663	38	17716.307	58	17718.145		
19	17714.724	39	17716.360	59	17718.225		
20	17714.804	40	17716.458	60	17718.328		

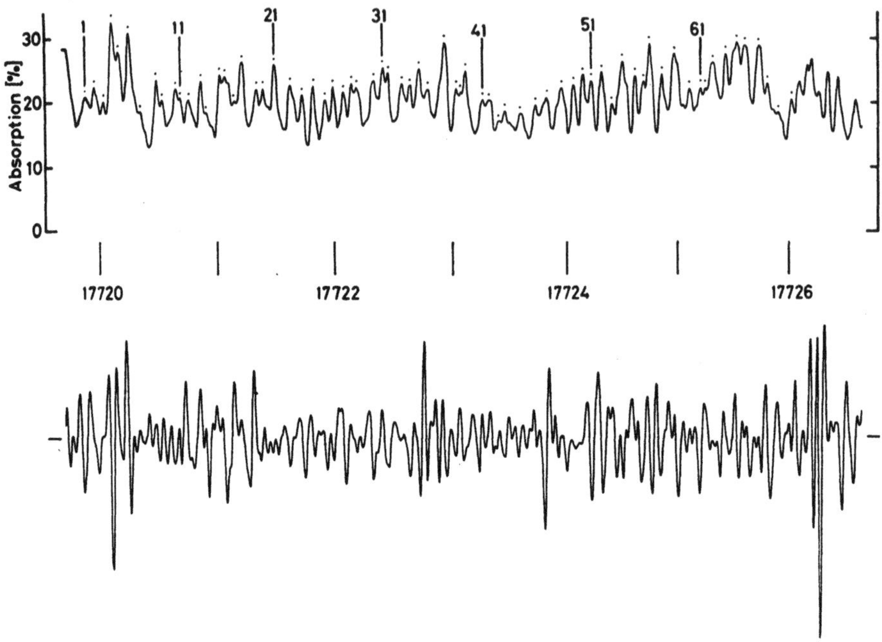

1	17719.898	21	17721.487	41	17723.266	61	17725.205
2	17719.965	22	17721.625	42	17723.332	62	17725.251
3	17720.043	23	17721.724	43	17723.408	63	17725.297
4	17720.121	24	17721.818	44	17723.462	64	17725.426
5	17720.171	25	17721.922	45	17723.600	65	17725.531
6	17720.255	26	17721.994	46	17723.726	66	17725.607
7	17720.344	27	17722.080	47	17723.821	67	17725.719
8	17720.485	28	17722.151	48	17723.962	68	17725.801
9	17720.528	29	17722.191	49	17724.062	69	17725.910
10	17720.643	30	17722.333	50	17724.147	70	17726.027
11	17720.687	31	17722.414	51	17724.225		
12	17720.753	32	17722.455	52	17724.316		
13	17720.862	33	17722.579	53	17724.408		
14	17720.912	34	17722.645	54	17724.504		
15	17721.024	35	17722.718	55	17724.619		
16	17721.071	36	17722.795	56	17724.690		
17	17721.153	37	17722.939	57	17724.751		
18	17721.217	38	17723.040	58	17724.860		
19	17721.336	39	17723.076	59	17724.972		
20	17721.392	40	17723.121	60	17725.107		

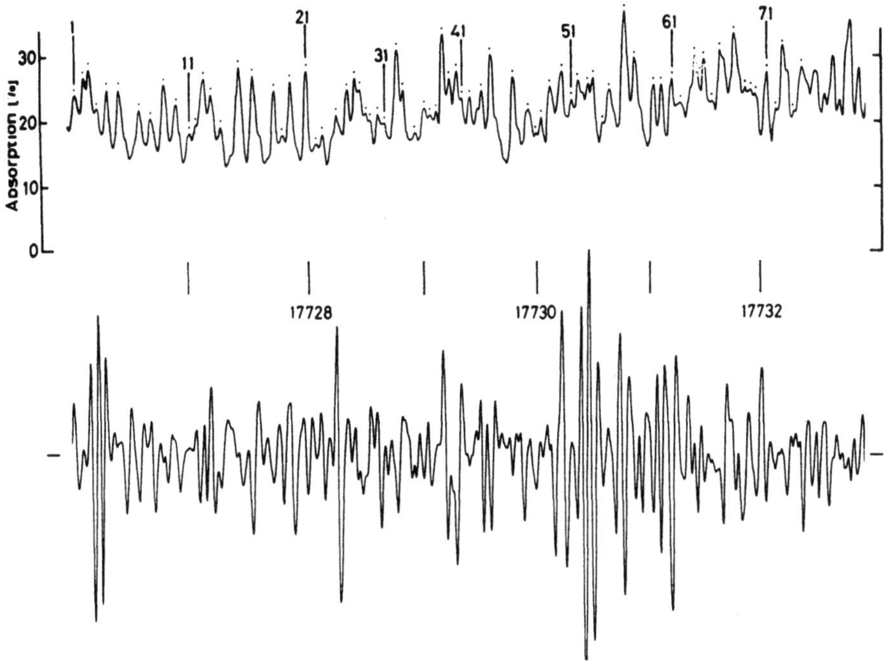

1	17726.093	21	17727.995	41	17729.350	61	17731.205
2	17726.157	22	17728.086	42	17729.421	62	17731.281
3	17726.203	23	17728.137	43	17729.520	63	17731.420
4	17726.273	24	17728.256	44	17729.602	64	17731.501
5	17726.350	25	17728.349	45	17729.795	65	17731.584
6	17726.441	26	17728.413	46	17729.936	66	17731.644
7	17726.606	27	17728.452	47	17729.997	67	17731.769
8	17726.702	28	17728.507	48	17730.048	68	17731.870
9	17726.816	29	17728.545	49	17730.126	69	17731.921
10	17726.919	30	17728.610	50	17730.225	70	17731.965
11	17727.025	31	17728.664	51	17730.314	71	17732.064
12	17727.083	32	17728.775	52	17730.369	72	17732.148
13	17727.146	33	17728.832	53	17730.460	73	17732.200
14	17727.207	34	17728.930	54	17730.506	74	17732.310
15	17727.287	35	17729.017	55	17730.594	75	17732.376
16	17727.437	36	17729.074	56	17730.646		
17	17727.552	37	17729.126	57	17730.776		
18	17727.730	38	17729.178	58	17730.869		
19	17727.799	39	17729.238	59	17731.040		
20	17727.866	40	17729.304	60	17731.112		

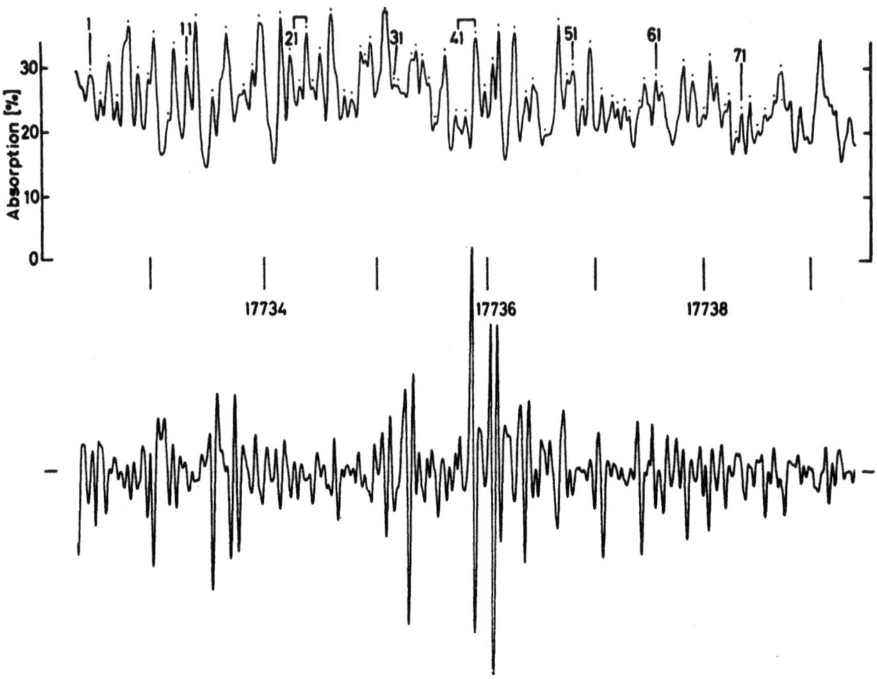

1	17732.498	21	17734.387	41	17735.888	61	17737.554
2	17732.582	22	17734.448	42	17735.973	62	17737.610
3	17732.659	23	17734.508	43	17736.054	63	17737.813
4	17732.725	24	17734.593	44	17736.104	64	17737.897
5	17732.818	25	17734.713	45	17736.250	65	17737.994
6	17732.904	26	17734.779	46	17736.360	66	17738.057
7	17732.995	27	17734.860	47	17736.430	67	17738.120
8	17733.044	28	17734.898	48	17736.547	68	17738.195
9	17733.169	29	17734.949	49	17736.663	69	17738.233
10	17733.218	30	17735.064	50	17736.749	70	17738.305
11	17733.333	31	17735.155	51	17736.797	71	17738.357
12	17733.414	32	17735.185	52	17736.882	72	17738.434
13	17733.563	33	17735.301	53	17736.957	73	17738.514
14	17733.688	34	17735.343	54	17737.060	74	17738.575
15	17733.835	35	17735.407	55	17737.158	75	17738.661
16	17733.911	36	17735.467	56	17737.211	76	17738.722
17	17733.978	37	17735.522	57	17737.271		
18	17734.157	38	17735.612	58	17737.315		
19	17734.242	39	17735.710	59	17737.408		
20	17734.329	40	17735.801	60	17737.453		

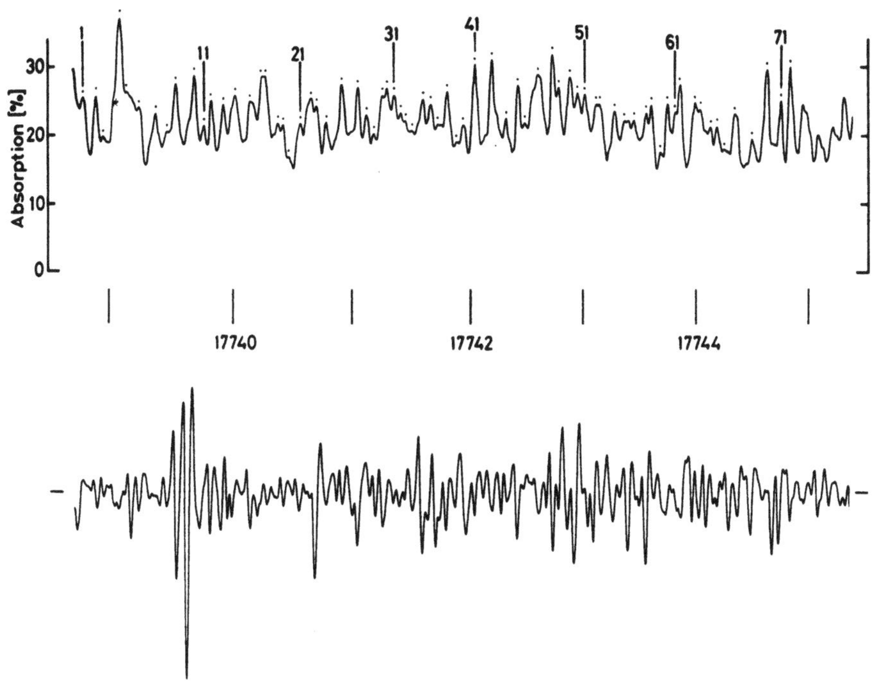

1	17738.810	21	17740.573	41	17742.047	61	17743.822
2	17738.909	22	17740.667	42	17742.190	62	17743.865
3	17738.967	23	17740.716	43	17742.319	63	17744.000
4	17739.093	24	17740.797	44	17742.420	64	17744.045
5	17739.152	25	17740.928	45	17742.482	65	17744.128
6	17739.246	26	17741.063	46	17742.602	66	17744.191
7	17739.382	27	17741.132	47	17742.724	67	17744.276
8	17739.471	28	17741.198	48	17742.784	68	17744.348
9	17739.544	29	17741.266	49	17742.884	69	17744.500
10	17739.694	30	17741.306	50	17742.953	70	17744.633
11	17739.767	31	17741.366	51	17743.020	71	17744.758
12	17739.830	32	17741.429	52	17743.122	72	17744.842
13	17739.927	33	17741.466	53	17743.157		
14	17740.025	34	17741.521	54	17743.284		
15	17740.160	35	17741.609	55	17743.382		
16	17740.250	36	17741.671	56	17743.460		
17	17740.279	37	17741.745	57	17743.566		
18	17740.381	38	17741.815	58	17743.613		
19	17740.428	39	17741.893	59	17743.699		
20	17740.478	40	17741.948	60	17743.754		

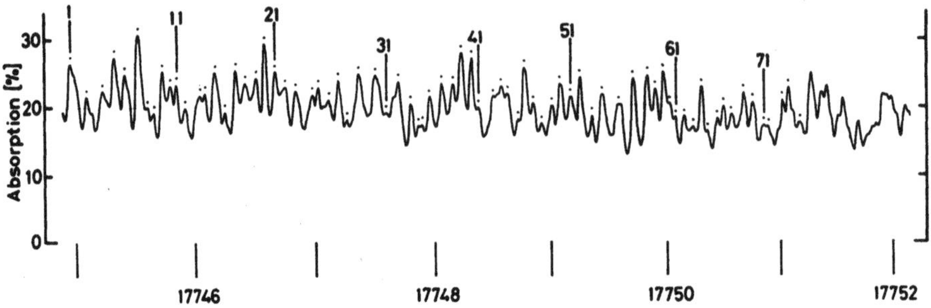

1	17744.957	21	17746.670	41	17748.370	61	17750.072
2	17745.091	22	17746.757	42	17748.502	62	17750.148
3	17745.228	23	17746.838	43	17748.571	63	17750.225
4	17745.327	24	17746.974	44	17748.625	64	17750.299
5	17745.421	25	17747.024	45	17748.717	65	17750.355
6	17745.537	26	17747.113	46	17748.770	66	17750.437
7	17745.609	27	17747.194	47	17748.842	67	17750.491
8	17745.662	28	17747.266	48	17748.916	68	17750.559
9	17745.738	29	17747.364	49	17749.003	69	17750.669
10	17745.801	30	17747.503	50	17749.072	70	17750.744
11	17745.851	31	17747.591	51	17749.166	71	17750.848
12	17745.925	32	17747.693	52	17749.247	72	17750.888
13	17746.040	33	17747.795	53	17749.348	73	17751.019
14	17746.087	34	17747.862	54	17749.437	74	17751.073
15	17746.173	35	17747.893	55	17749.581	75	17751.168
16	17746.255	36	17747.958	56	17749.704		
17	17746.342	37	17748.059	57	17749.827		
18	17746.422	38	17748.157	58	17749.897		
19	17746.513	39	17748.225	59	17749.968		
20	17746.583	40	17748.314	60	17750.023		

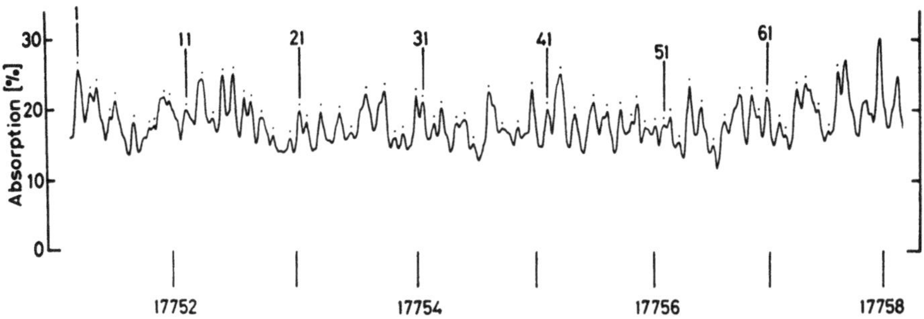

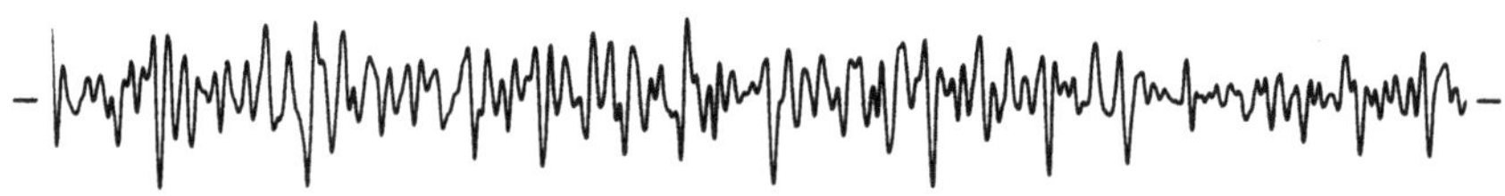

1	17751.279	21	17753.034	41	17755.099	61	17756.980
2	17751.364	22	17753.094	42	17755.211	62	17757.091
3	17751.412	23	17753.212	43	17755.327	63	17757.143
4	17751.508	24	17753.371	44	17755.487	64	17757.236
5	17751.552	25	17753.466	45	17755.603	65	17757.311
6	17751.690	26	17753.592	46	17755.716	66	17757.419
7	17751.818	27	17753.733	47	17755.814	67	17757.516
8	17751.853	28	17753.817	48	17755.858	68	17757.601
9	17751.938	29	17753.884	49	17755.926		
10	17751.983	30	17753.994	50	17756.013		
11	17752.114	31	17754.049	51	17756.085		
12	17752.237	32	17754.142	52	17756.140		
13	17752.318	33	17754.203	53	17756.218		
14	17752.400	34	17754.325	54	17756.309		
15	17752.485	35	17754.397	55	17756.408		
16	17752.579	36	17754.464	56	17756.511		
17	17752.631	37	17754.603	57	17756.612		
18	17752.717	38	17754.709	58	17756.740		
19	17752.818	39	17754.848	59	17756.853		
20	17752.960	40	17754.970	60	17756.907		

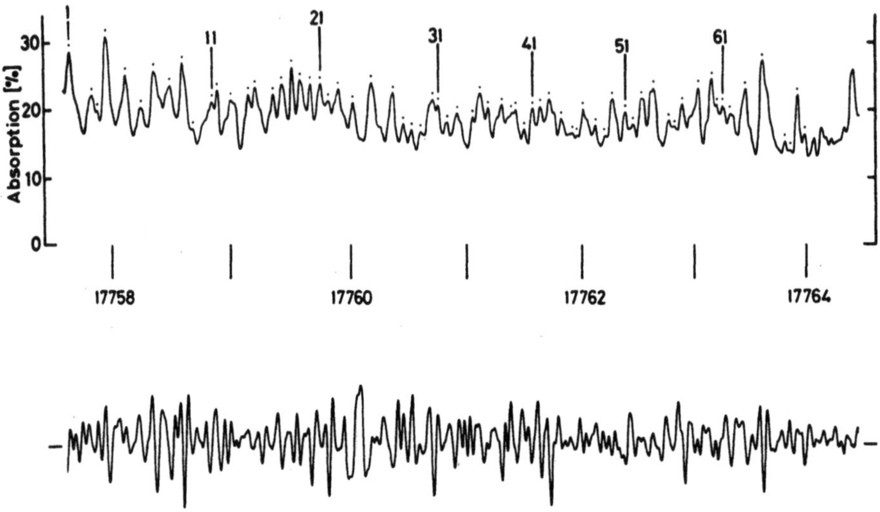

1	17757.671	21	17759.750	41	17761.574	61	17763.252
2	17757.838	22	17759.814	42	17761.639	62	17763.327
3	17757.886	23	17759.902	43	17761.718	63	17763.458
4	17757.966	24	17760.024	44	17761.823	64	17763.610
5	17758.121	25	17760.182	45	17761.920	65	17763.817
6	17758.245	26	17760.365	46	17761.964	66	17763.863
7	17758.355	27	17760.455	47	17762.016	67	17763.929
8	17758.478	28	17760.527	48	17762.126	68	17763.991
9	17758.584	29	17760.606	49	17762.209		
10	17758.671	30	17760.711	50	17762.273		
11	17758.834	31	17760.755	51	17762.383		
12	17758.883	32	17760.834	52	17762.453		
13	17759.000	33	17760.916	53	17762.523		
14	17759.140	34	17761.052	54	17762.640		
15	17759.197	35	17761.114	55	17762.767		
16	17759.346	36	17761.184	56	17762.829		
17	17759.419	37	17761.304	57	17762.894		
18	17759.504	38	17761.391	58	17763.039		
19	17759.573	39	17761.424	59	17763.158		
20	17759.664	40	17761.500	60	17763.206		

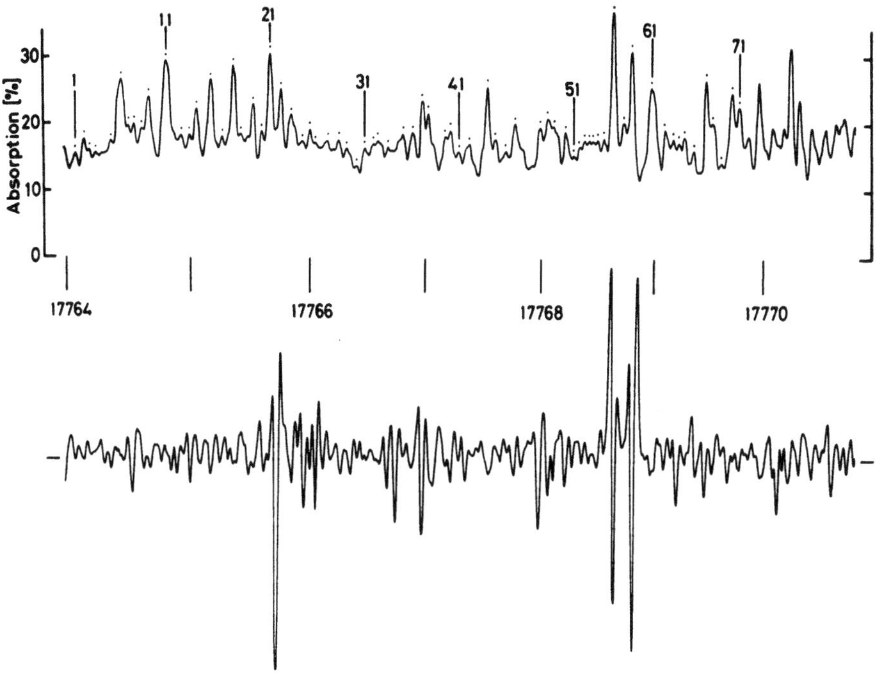

1	17764.077	21	17765.672	41	17767.289	61	17768.988
2	17764.145	22	17765.767	42	17767.382	62	17769.107
3	17764.186	23	17765.848	43	17767.546	63	17769.181
4	17764.235	24	17765.942	44	17767.608	64	17769.227
5	17764.358	25	17766.010	45	17767.694	65	17769.280
6	17764.439	26	17766.055	46	17767.783	66	17769.366
7	17764.496	27	17766.150	47	17767.996	67	17769.485
8	17764.541	28	17766.253	48	17768.063	68	17769.542
9	17764.602	29	17766.320	49	17768.114	69	17769.616
10	17764.662	30	17766.411	50	17768.217	70	17769.726
11	17764.798	31	17766.479	51	17768.292	71	17769.786
12	17764.925	32	17766.552	52	17768.341		
13	17764.994	33	17766.593	53	17768.387		
14	17765.050	34	17766.682	54	17768.427		
15	17765.168	35	17766.811	55	17768.465		
16	17765.268	36	17766.892	56	17768.503		
17	17765.357	37	17766.979	57	17768.564		
18	17765.426	38	17767.031	58	17768.640		
19	17765.525	39	17767.178	59	17768.735		
20	17765.598	40	17767.218	60	17768.809		

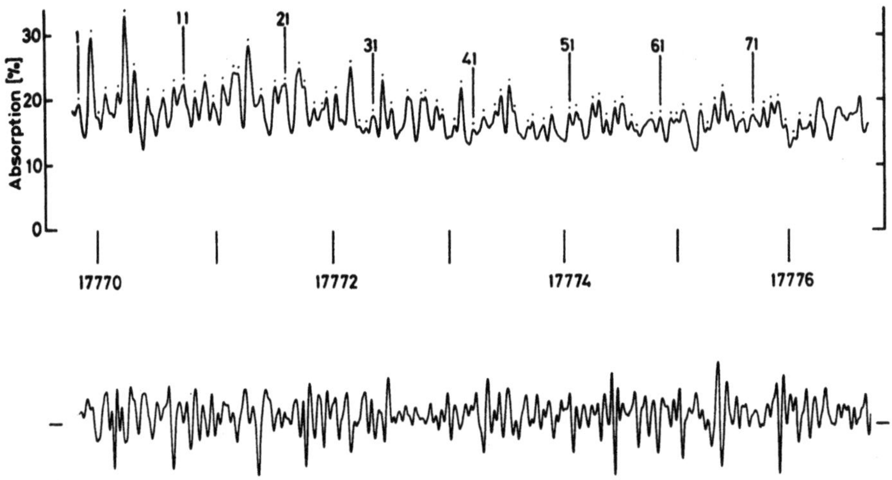

1	17769.865	21	17771.603	41	17773.206	61	17774.848
2	17769.964	22	17771.725	42	17773.297	62	17774.942
3	17770.019	23	17771.767	43	17773.398	63	17774.994
4	17770.080	24	17771.850	44	17773.446	64	17775.056
5	17770.183	25	17771.921	45	17773.518	65	17775.213
6	17770.249	26	17771.957	46	17773.561	66	17775.274
7	17770.325	27	17772.038	47	17773.681	67	17775.337
8	17770.432	28	17772.161	48	17773.719	68	17775.411
9	17770.562	29	17772.233	49	17773.815	69	17775.481
10	17770.655	30	17772.290	50	17773.890	70	17775.590
11	17770.733	31	17772.349	51	17774.045	71	17775.677
12	17770.829	32	17772.431	52	17774.104	72	17775.779
13	17770.920	33	17772.507	53	17774.247	73	17775.848
14	17770.985	34	17772.647	54	17774.302	74	17775.910
15	17771.072	35	17772.764	55	17774.366	75	17775.975
16	17771.164	36	17772.796	56	17774.441	76	17776.049
17	17771.204	37	17772.892	57	17774.509	77	17776.101
18	17771.296	38	17772.939	58	17774.583	78	17776.197
19	17771.400	39	17773.042	59	17774.629		
20	17771.520	40	17773.102	60	17774.776		

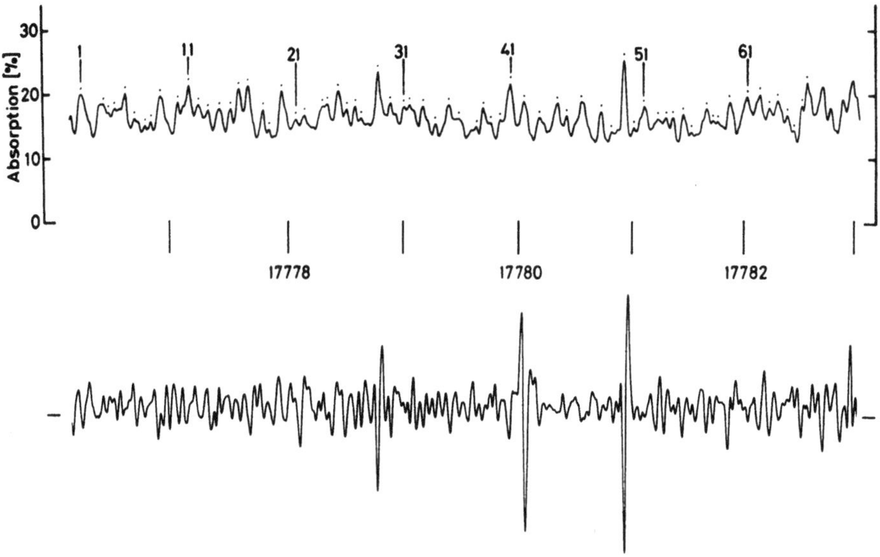

1	17776.281	21	17778.078	41	17779.932	61	17782.036
2	17776.463	22	17778.146	42	17780.050	62	17782.149
3	17776.504	23	17778.300	43	17780.189	63	17782.235
4	17776.552	24	17778.340	44	17780.347	64	17782.309
5	17776.645	25	17778.434	45	17780.426	65	17782.393
6	17776.723	26	17778.509	46	17780.565	66	17782.456
7	17776.808	27	17778.582	47	17780.733	67	17782.528
8	17776.859	28	17778.638	48	17780.834	68	17782.581
9	17776.941	29	17778.786	49	17780.936		
10	17777.084	30	17778.897	50	17781.020		
11	17777.172	31	17779.012	51	17781.116		
12	17777.254	32	17779.065	52	17781.235		
13	17777.331	33	17779.182	53	17781.305		
14	17777.422	34	17779.290	54	17781.364		
15	17777.516	35	17779.406	55	17781.459		
16	17777.585	36	17779.477	56	17781.535		
17	17777.664	37	17779.636	57	17781.674		
18	17777.794	38	17779.701	58	17781.732		
19	17777.847	39	17779.759	59	17781.780		
20	17777.953	40	17779.840	60	17781.883		

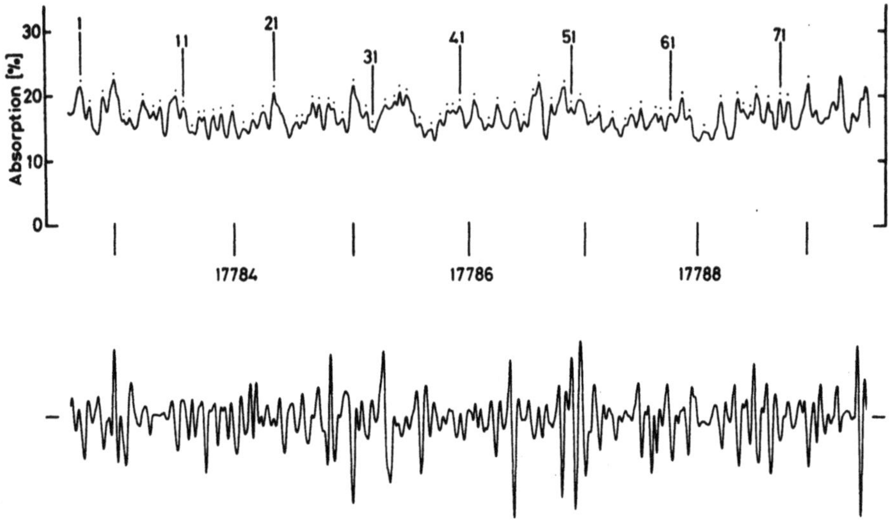

1	17782.722	21	17784.344	41	17785.926	61	17787.768
2	17782.790	22	17784.532	42	17786.047	62	17787.868
3	17782.903	23	17784.588	43	17786.167	63	17787.929
4	17782.997	24	17784.666	44	17786.244	64	17788.086
5	17783.083	25	17784.722	45	17786.401	65	17788.219
6	17783.127	26	17784.796	46	17786.481	66	17788.375
7	17783.234	27	17784.838	47	17786.558	67	17788.428
8	17783.326	28	17784.912	48	17786.607	68	17788.496
9	17783.379	29	17785.007	49	17786.715	69	17788.556
10	17783.506	30	17785.111	50	17786.826	70	17788.659
11	17783.576	31	17785.167	51	17786.885	71	17788.767
12	17783.651	32	17785.272	52	17786.970	72	17788.838
13	17783.707	33	17785.363	53	17787.054	73	17789.015
14	17783.744	34	17785.403	54	17787.132		
15	17783.829	35	17785.458	55	17787.254		
16	17783.899	36	17785.575	56	17787.365		
17	17783.996	37	17785.662	57	17787.415		
18	17784.082	38	17785.749	58	17787.502		
19	17784.165	39	17785.815	59	17787.633		
20	17784.247	40	17785.858	60	17787.684		

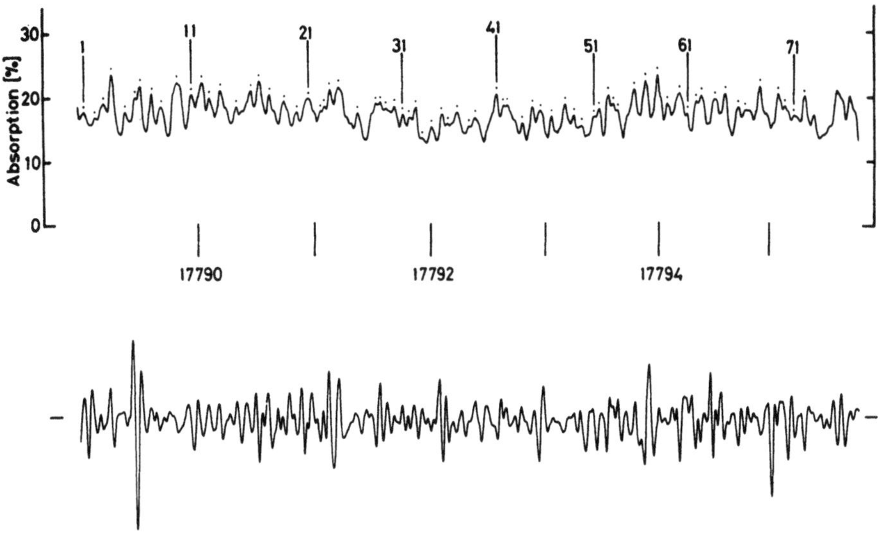

1	17789.082	21	17790.956	41	17792.577	61	17794.254
2	17789.172	22	17791.061	42	17792.638	62	17794.339
3	17789.230	23	17791.101	43	17792.676	63	17794.387
4	17789.299	24	17791.147	44	17792.803	64	17794.513
5	17789.411	25	17791.223	45	17792.890	65	17794.599
6	17789.492	26	17791.382	46	17792.959	66	17794.719
7	17789.533	27	17791.533	47	17793.062	67	17794.787
8	17789.629	28	17791.575	48	17793.180	68	17794.917
9	17789.707	29	17791.623	49	17793.253	69	17795.092
10	17789.840	30	17791.691	50	17793.315	70	17795.150
11	17789.959	31	17791.764	51	17793.425	71	17795.247
12	17790.046	32	17791.822	52	17793.477	72	17795.342
13	17790.110	33	17791.870	53	17793.558		
14	17790.201	34	17791.928	54	17793.604		
15	17790.337	35	17792.008	55	17793.646		
16	17790.452	36	17792.099	56	17793.785		
17	17790.528	37	17792.142	57	17793.889		
18	17790.618	38	17792.235	58	17793.992		
19	17790.750	39	17792.336	59	17794.079		
20	17790.856	40	17792.393	60	17794.189		

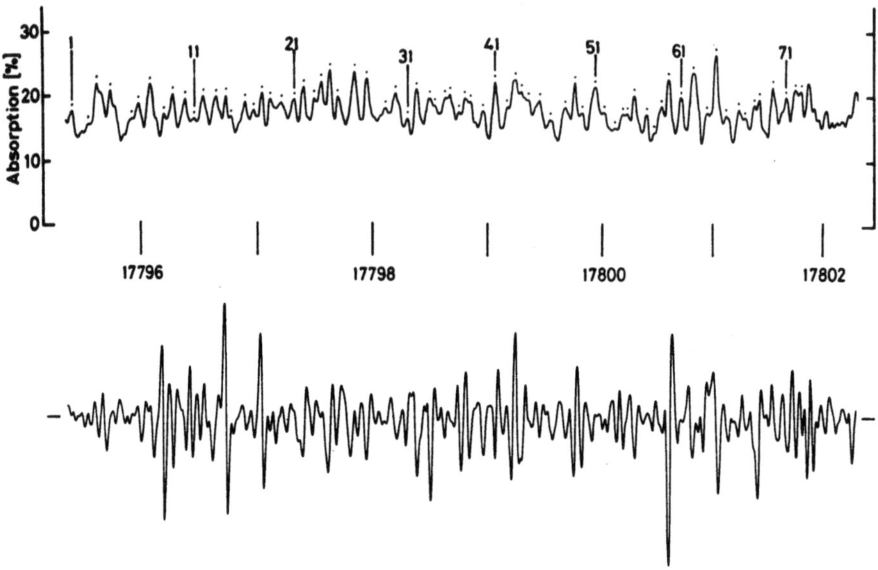

1	17795.426	21	17797.324	41	17799.065	61	17800.714
2	17795.560	22	17797.411	42	17799.173	62	17800.827
3	17795.635	23	17797.505	43	17799.248	63	17800.957
4	17795.752	24	17797.568	44	17799.293	64	17801.034
5	17795.931	25	17797.640	45	17799.355	65	17801.126
6	17795.986	26	17797.713	46	17799.452	66	17801.244
7	17796.083	27	17797.854	47	17799.543	67	17801.300
8	17796.199	28	17797.960	48	17799.682	68	17801.389
9	17796.275	29	17798.108	49	17799.765	69	17801.430
10	17796.379	30	17798.198	50	17799.831	70	17801.559
11	17796.453	31	17798.302	51	17799.944	71	17801.683
12	17796.537	32	17798.379	52	17800.018	72	17801.761
13	17796.642	33	17798.492	53	17800.119	73	17801.812
14	17796.733	34	17798.538	54	17800.192		
15	17796.783	35	17798.632	55	17800.229		
16	17796.897	36	17798.668	56	17800.294		
17	17796.971	37	17798.749	57	17800.406		
18	17797.046	38	17798.802	58	17800.474		
19	17797.119	39	17798.851	59	17800.541		
20	17797.210	40	17798.960	60	17800.603		

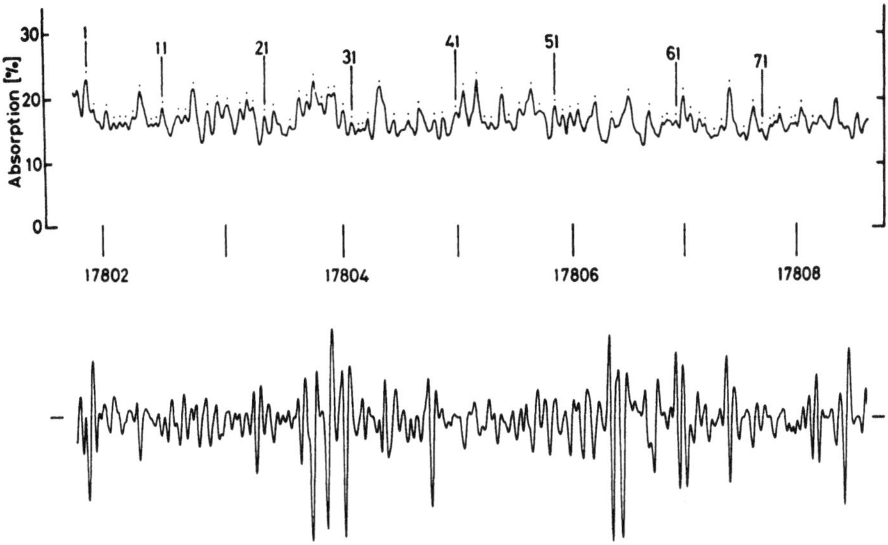

1	17801.877	21	17803.334	41	17804.988	61	17806.938
2	17801.934	22	17803.416	42	17805.051	62	17806.997
3	17802.038	23	17803.547	43	17805.163	63	17807.060
4	17802.096	24	17803.631	44	17805.238	64	17807.140
5	17802.142	25	17803.696	45	17805.296	65	17807.193
6	17802.191	26	17803.755	46	17805.382	66	17807.279
7	17802.248	27	17803.830	47	17805.448	67	17807.337
8	17802.310	28	17803.885	48	17805.535	68	17007.407
9	17802.399	29	17803.930	49	17805.641	69	17807.527
10	17802.446	30	17804.010	50	17805.725	70	17807.619
11	17802.495	31	17804.086	51	17805.854	71	17807.688
12	17802.628	32	17804.148	52	17805.925	72	17807.765
13	17802.688	33	17804.186	53	17805.986	73	17807.805
14	17802.752	34	17804.227	54	17806.059	74	17807.854
15	17802.863	35	17804.322	55	17806.211	75	17807.957
16	17802.949	36	17804.452	56	17806.356	76	17807.992
17	17803.024	37	17804.580	57	17806.506	77	17808.049
18	17803.135	38	17804.675	58	17806.691	78	17808.162
19	17803.192	39	17804.798	59	17806.806		
20	17803.242	40	17804.875	60	17806.855		

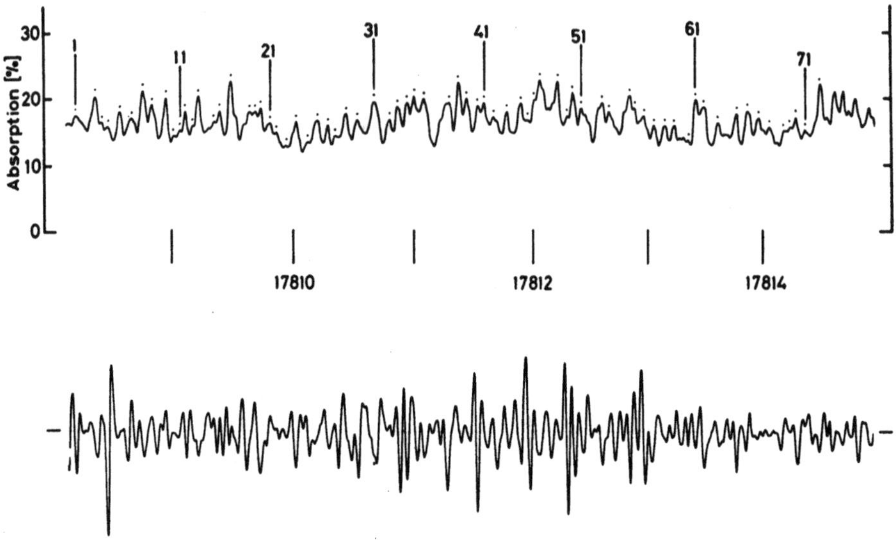

1	17808.235	21	17809.819	41	17811.588	61	17813.425
2	17808.382	22	17809.876	42	17811.667	62	17813.495
3	17808.436	23	17809.950	43	17811.780	63	17813.619
4	17808.488	24	17810.034	44	17811.898	64	17813.779
5	17808.587	25	17810.129	45	17811.953	65	17813.876
6	17808.694	26	17810.212	46	17812.016	66	17813.967
7	17808.775	27	17810.300	47	17812.061	67	17814.063
8	17808.846	28	17810.360	48	17812.215	68	17814.189
9	17808.966	29	17810.446	49	17812.297	69	17814.241
10	17809.037	30	17810.538	50	17812.345	70	17814.289
11	17809.081	31	17810.674	51	17812.418	71	17814.367
12	17809.126	32	17810.802	52	17812.600	72	17814.493
13	17809.187	33	17810.866	53	17812.659		
14	17809.239	34	17810.946	54	17812.842		
15	17809.370	35	17811.001	55	17812.884		
16	17809.416	36	17811.081	56	17812.967		
17	17809.508	37	17811.290	57	17813.056		
18	17809.660	38	17811.369	58	17813.150		
19	17809.699	39	17811.439	59	17813.238		
20	17809.746	40	17811.534	60	17813.357		

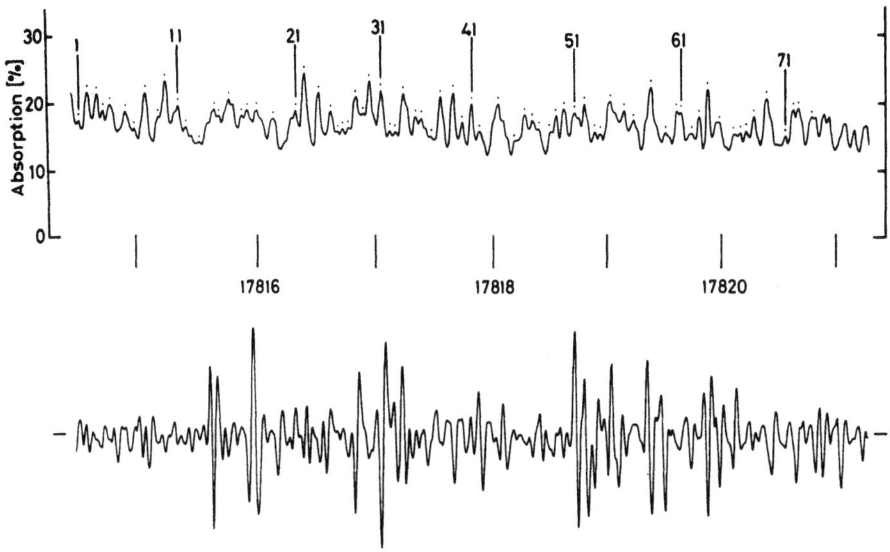

1	17814.558	21	17816.341	41	17817.833	61	17819.652
2	17814.623	22	17816.417	42	17817.892	62	17819.749
3	17814.697	23	17816.531	43	17818.060	63	17819.812
4	17814.747	24	17816.629	44	17818.207	64	17819.886
5	17814.811	25	17816.687	45	17818.296	65	17819.990
6	17814.928	26	17816.729	46	17818.365	66	17820.118
7	17815.002	27	17816.780	47	17818.416	67	17820.170
8	17815.090	28	17816.839	48	17818.523	68	17820.227
9	17815.197	29	17816.899	49	17818.575	69	17820.296
10	17815.254	30	17816.953	50	17818.643	70	17820.397
11	17815.359	31	17817.051	51	17818.739	71	17820.570
12	17815.424	32	17817.122	52	17818.822	72	17820.632
13	17815.536	33	17817.171	53	17818.909	73	17820.683
14	17815.657	34	17817.237	54	17818.957	74	17820.800
15	17815.699	35	17817.339	55	17819.040		
16	17815.786	36	17817.393	56	17819.162		
17	17815.891	37	17817.481	57	17819.240		
18	17815.936	38	17817.561	58	17819.386		
19	17816.014	39	17817.670	59	17819.509		
20	17816.166	40	17817.748	60	17819.611		

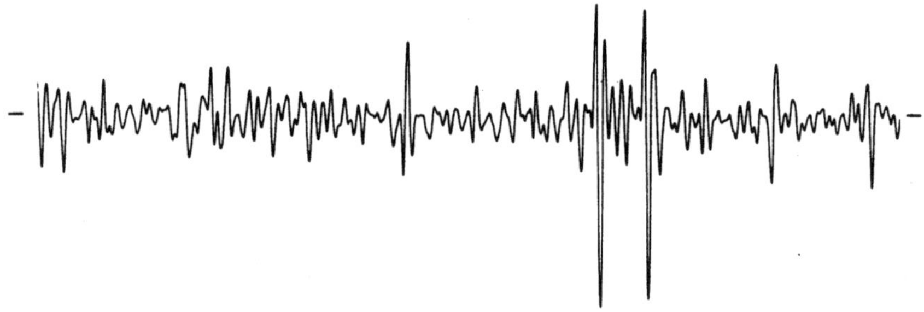

1	17820.905	21	17822.783	41	17824.678	61	17826.672
2	17820.959	22	17822.916	42	17824.755	62	17826.715
3	17821.091	23	17822.987	43	17824.862	63	17826.810
4	17821.181	24	17823.109	44	17825.030	64	17826.932
5	17821.276	25	17823.187	45	17825.169	65	17827.044
6	17821.383	26	17823.235	46	17825.223	66	17827.182
7	17821.431	27	17823.303	47	17825.284		
8	17821.532	28	17823.358	48	17825.335		
9	17821.604	29	17823.414	49	17825.430		
10	17821.650	30	17823.568	50	17825.571		
11	17821.745	31	17823.664	51	17825.664		
12	17821.846	32	17823.818	52	17825.776		
13	17821.986	33	17823.892	53	17825.880		
14	17822.032	34	17823.992	54	17826.029		
15	17822.077	35	17824.113	55	17826.078		
16	17822.252	36	17824.205	56	17826.219		
17	17822.339	37	17824.257	57	17826.310		
18	17822.439	38	17824.365	58	17826.411		
19	17822.588	39	17824.432	59	17826.481		
20	17822.685	40	17824.584	60	17826.559		

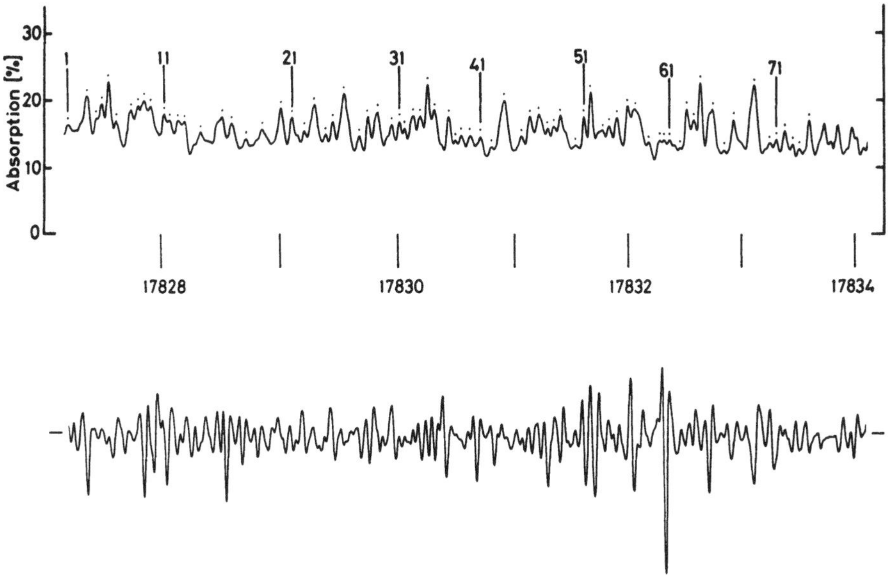

1	17827.249	21	17829.116	41	17830.721	61	17832.362
2	17827.410	22	17829.219	42	17830.809	62	17832.454
3	17827.489	23	17829.308	43	17830.921	63	17832.511
4	17827.529	24	17829.400	44	17831.060	64	17832.577
5	17827.589	25	17829.460	45	17831.133	65	17832.641
6	17827.643	26	17829.557	46	17831.212	66	17832.747
7	17827.771	27	17829.682	47	17831.288	67	17832.847
8	17827.820	28	17829.757	48	17831.341	68	17832.931
9	17827.882	29	17829.834	49	17831.399	69	17833.115
10	17827.933	30	17829.957	50	17831.525	70	17833.244
11	17828.047	31	17830.024	51	17831.609	71	17833.302
12	17828.090	32	17830.069	52	17831.669	72	17833.379
13	17828.162	33	17830.146	53	17831.768	73	17833.447
14	17828.217	34	17830.209	54	17831.831	74	17833.509
15	17828.348	35	17830.271	55	17831.901	75	17833.601
16	17828.525	36	17830.328	56	17832.000		
17	17828.608	37	17830.455	57	17832.053		
18	17828.726	38	17830.511	58	17832.180		
19	17828.863	39	17830.563	59	17832.276		
20	17829.025	40	17830.637	60	17832.318		

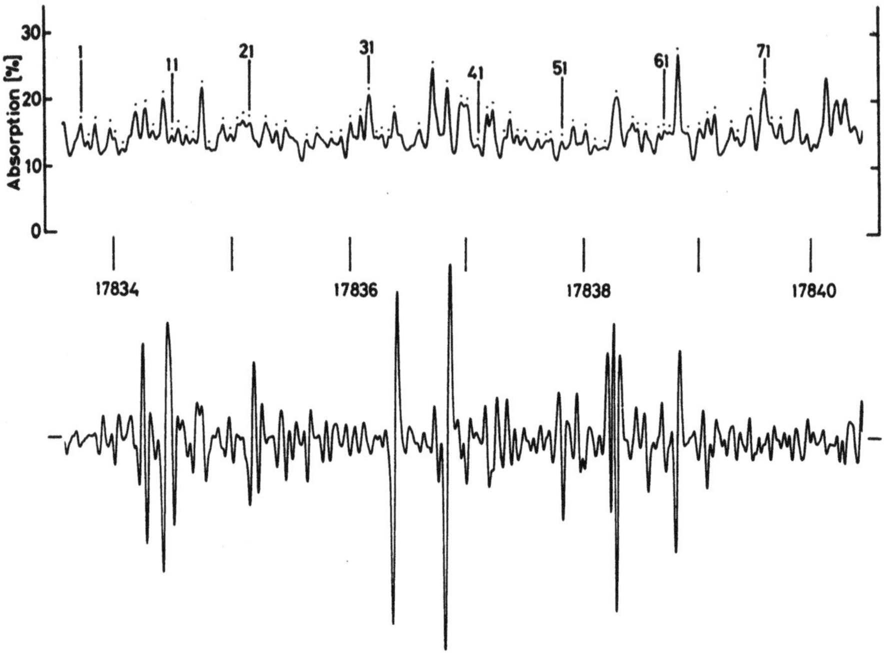

1	17833.735	21	17835.155	41	17837.104	61	17838.710
2	17833.787	22	17835.288	42	17837.187	62	17838.754
3	17833.851	23	17835.379	43	17837.234	63	17838.830
4	17833.976	24	17835.455	44	17837.335	64	17838.914
5	17834.020	25	17835.640	45	17837.381	65	17839.017
6	17834.080	26	17835.726	46	17837.438	66	17839.089
7	17834.191	27	17835.850	47	17837.506	67	17839.149
8	17834.268	28	17835.931	48	17837.612	68	17839.300
9	17834.334	29	17836.013	49	17837.677	69	17839.354
10	17834.422	30	17836.099	50	17837.722	70	17839.476
11	17834.493	31	17836.176	51	17837.815	71	17839.601
12	17834.545	32	17836.234	52	17837.911	72	17839.654
13	17834.613	33	17836.284	53	17838.019	73	17839.738
14	17834.668	34	17836.345	54	17838.096		
15	17834.747	35	17836.392	55	17838.182		
16	17834.805	36	17836.607	56	17838.291		
17	17834.930	37	17836.720	57	17838.425		
18	17834.995	38	17836.843	58	17838.476		
19	17835.055	39	17836.965	59	17838.542		
20	17835.096	40	17837.012	60	17838.660		

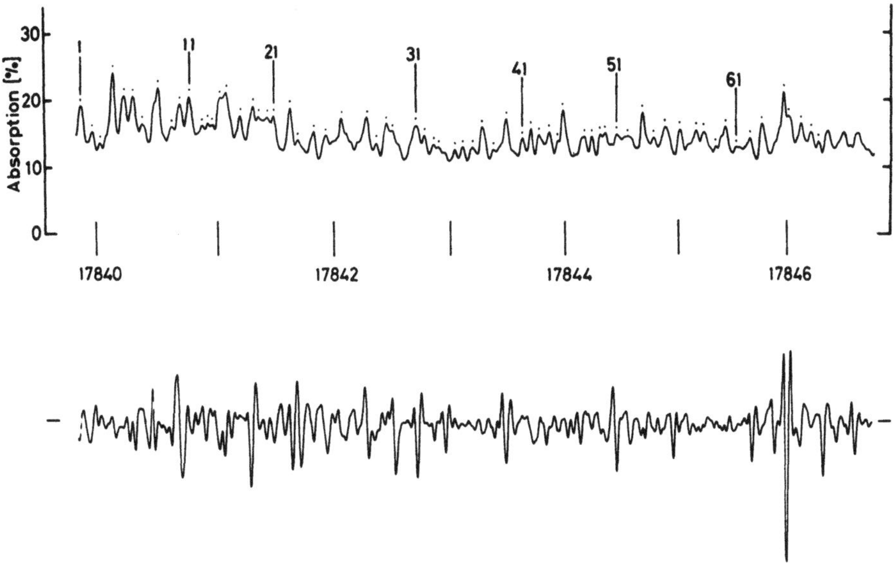

1	17839.886	21	17841.489	41	17843.631	61	17845.532
2	17839.974	22	17841.635	42	17843.708	62	17845.659
3	17840.039	23	17841.705	43	17843.777	63	17845.769
4	17840.148	24	17841.835	44	17843.870	64	17845.975
5	17840.241	25	17841.942	45	17843.945	65	17846.026
6	17840.313	26	17842.075	46	17843.992	66	17846.143
7	17840.394	27	17842.286	47	17844.174	67	17846.230
8	17840.519	28	17842.369	48	17844.246	68	17846.299
9	17840.636	29	17842.453	49	17844.324		
10	17840.697	30	17842.508	50	17844.366		
11	17840.776	31	17842.714	51	17844.471		
12	17840.884	32	17842.781	52	17844.564		
13	17840.932	33	17842.861	53	17844.698		
14	17840.972	34	17842.905	54	17844.794		
15	17841.030	35	17843.045	55	17844.895		
16	17841.082	36	17843.112	56	17845.023		
17	17841.202	37	17843.196	57	17845.165		
18	17841.309	38	17843.265	58	17845.239		
19	17841.362	39	17843.379	59	17845.343		
20	17841.436	40	17843.492	60	17845.439		

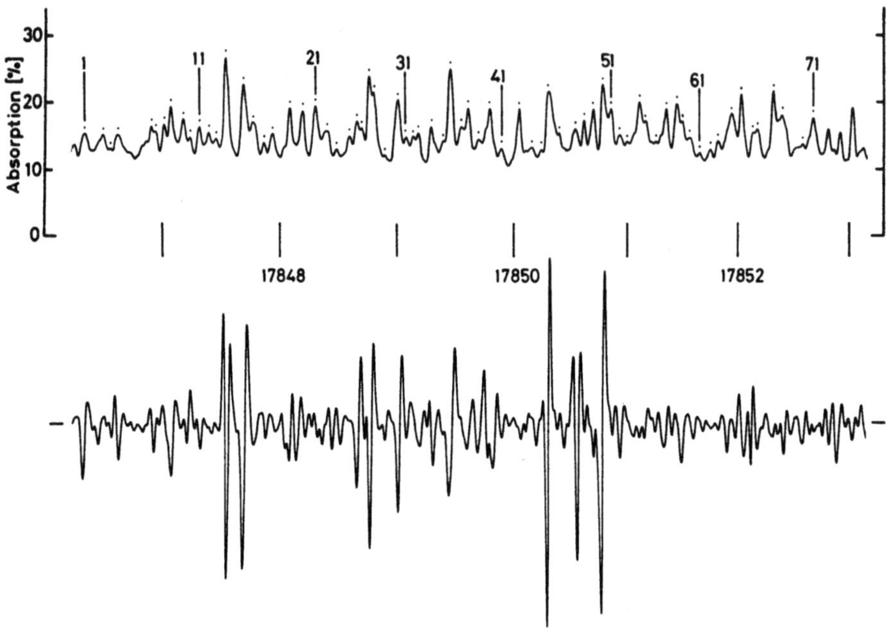

1	17846.384	21	17848.322	41	17849.904	61	17851.663
2	17846.534	22	17848.422	42	17850.059	62	17851.758
3	17846.590	23	17848.504	43	17850.168	63	17851.830
4	17846.657	24	17848.612	44	17850.254	64	17851.952
5	17846.929	25	17848.674	45	17850.319	65	17852.039
6	17846.966	26	17848.717	46	17850.408	66	17852.138
7	17847.030	27	17848.784	47	17850.556	67	17852.177
8	17847.090	28	17848.819	48	17850.627	68	17852.318
9	17847.191	29	17848.912	49	17850.709	69	17852.397
10	17847.242	30	17849.023	50	17850.791	70	17852.580
11	17847.322	31	17849.089	51	17850.865	71	17852.672
12	17847.403	32	17849.152	52	17850.940		
13	17847.469	33	17849.193	53	17851.007		
14	17847.563	34	17849.302	54	17851.120		
15	17847.708	35	17849.405	55	17851.178		
16	17847.787	36	17849.468	56	17851.266		
17	17847.875	37	17849.564	57	17851.363		
18	17847.949	38	17849.621	58	17851.461		
19	17848.097	39	17849.705	59	17851.511		
20	17848.210	40	17849.805	60	17851.573		

1	17852.806	21	17854.707	41	17856.732	61	17858.582
2	17852.860	22	17854.775	42	17856.845	62	17858.646
3	17852.914	23	17854.884	43	17856.904	63	17858.774
4	17853.029	24	17855.008	44	17856.964	64	17858.909
5	17853.125	25	17855.109	45	17857.062	65	17858.967
6	17853.230	26	17855.220	46	17857.118	66	17859.037
7	17853.353	27	17855.311	47	17857.161		
8	17853.411	28	17855.484	48	17857.283		
9	17853.503	29	17855.557	49	17857.415		
10	17853.596	30	17855.628	50	17857.543		
11	17853.708	31	17855.709	51	17857.614		
12	17853.817	32	17855.781	52	17857.765		
13	17853.886	33	17855.949	53	17857.849		
14	17854.053	34	17856.167	54	17858.006		
15	17854.101	35	17856.259	55	17858.065		
16	17854.171	36	17856.377	56	17858.122		
17	17854.278	37	17856.454	57	17858.177		
18	17854.333	38	17856.516	58	17858.274		
19	17854.410	39	17856.564	59	17858.350		
20	17854.505	40	17856.633	60	17858.457		

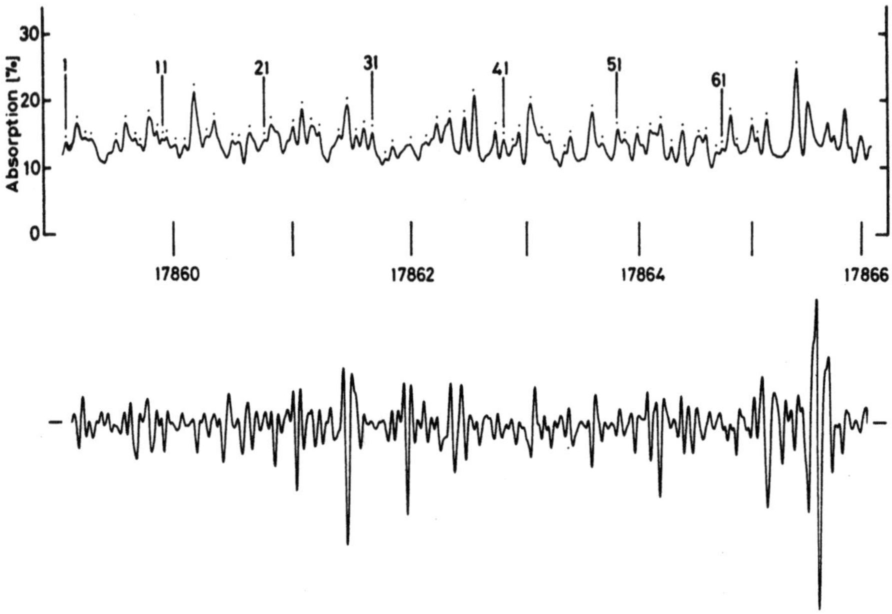

1	17859.111	21	17860.780	41	17862.802	61	17864.737
2	17859.197	22	17860.840	42	17862.888	62	17864.805
3	17859.266	23	17861.013	43	17862.937	63	17864.868
4	17859.316	24	17861.091	44	17863.040	64	17865.001
5	17859.526	25	17861.168	45	17863.136	65	17865.048
6	17859.606	26	17861.233	46	17863.208	66	17865.130
7	17859.685	27	17861.401	47	17863.339	67	17865.403
8	17859.730	28	17861.471	48	17863.390		
9	17859.810	29	17861.546	49	17863.587		
10	17859.862	30	17861.612	50	17863.672		
11	17859.914	31	17861.683	51	17863.810		
12	17859.946	32	17861.793	52	17863.869		
13	17860.027	33	17861.852	53	17863.984		
14	17860.106	34	17861.998	54	17864.097		
15	17860.187	35	17862.134	55	17864.189		
16	17860.295	36	17862.225	56	17864.288		
17	17860.360	37	17862.337	57	17864.386		
18	17860.515	38	17862.465	58	17864.530		
19	17860.571	39	17862.552	59	17864.588		
20	17860.657	40	17862.730	60	17864.687		

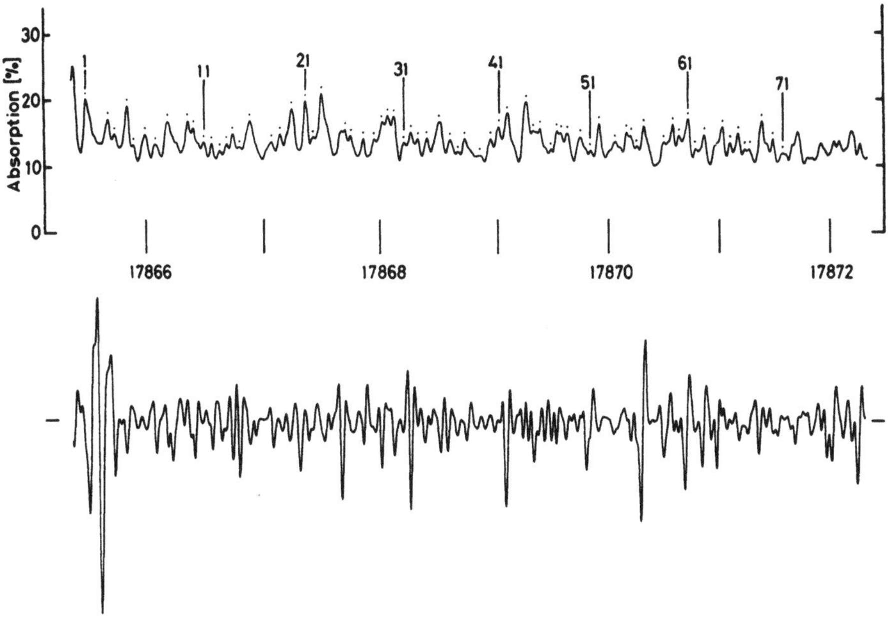

1	17865.506	21	17867.364	41	17869.010	61	17870.717
2	17865.688	22	17867.427	42	17869.091	62	17870.794
3	17865.742	23	17867.503	43	17869.258	63	17870.869
4	17865.846	24	17867.702	44	17869.321	64	17871.027
5	17865.900	25	17867.746	45	17869.381	65	17871.097
6	17865.996	26	17867.862	46	17869.483	66	17871.173
7	17866.083	27	17867.959	47	17869.532	67	17871.233
8	17866.187	28	17868.026	48	17869.574	68	17871.277
9	17866.358	29	17868.073	49	17869.630	69	17871.383
10	17866.400	30	17868.126	50	17869.747	70	17871.482
11	17866.493	31	17868.215	51	17869.846	71	17871.580
12	17866.559	32	17868.267	52	17869.920		
13	17866.627	33	17868.330	53	17870.066		
14	17866.687	34	17868.408	54	17870.171		
15	17866.736	35	17868.514	55	17870.206		
16	17866.797	36	17868.604	56	17870.262		
17	17866.884	37	17868.673	57	17870.323		
18	17867.072	38	17868.727	58	17870.497		
19	17867.147	39	17868.853	59	17870.576		
20	17867.243	40	17868.948	60	17870.634		

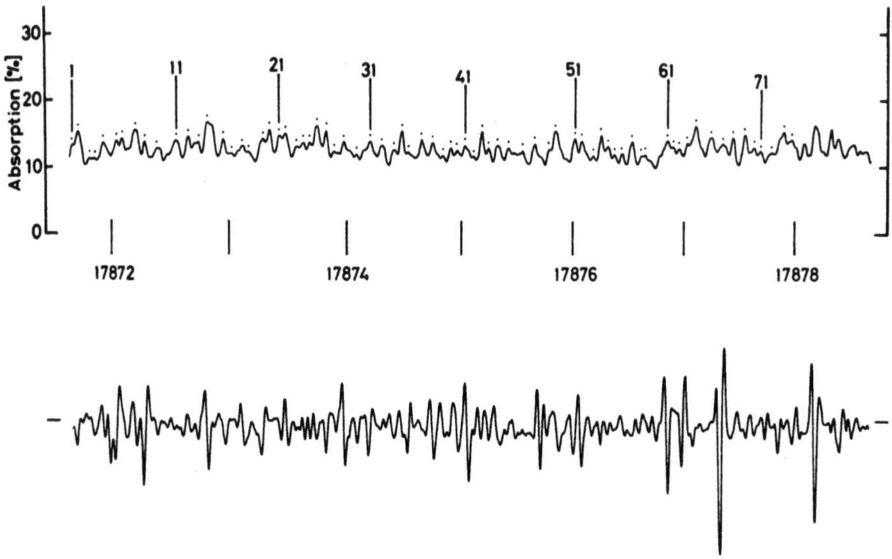

1	17871.670	21	17873.428	41	17875.040	61	17876.851
2	17871.715	22	17873.476	42	17875.118	62	17876.908
3	17871.806	23	17873.568	43	17875.193	63	17876.960
4	17871.847	24	17873.627	44	17875.249	64	17877.035
5	17871.927	25	17873.683	45	17875.328	65	17877.116
6	17872.042	26	17873.747	46	17875.425	66	17877.258
7	17872.093	27	17873.828	47	17875.553	67	17877.359
8	17872.204	28	17873.898	48	17875.664	68	17877.448
9	17872.284	29	17873.981	49	17875.789	69	17877.558
10	17872.382	30	17874.088	50	17875.848	70	17877.639
11	17872.551	31	17874.207	51	17876.023	71	17877.707
12	17872.651	32	17874.311	52	17876.081	72	17877.790
13	17872.732	33	17874.416	53	17876.185	73	17877.905
14	17872.815	34	17874.485	54	17876.257	74	17877.979
15	17872.954	35	17874.535	55	17876.316		
16	17873.028	36	17874.655	56	17876.374		
17	17873.120	37	17874.753	57	17876.443		
18	17873.176	38	17874.846	58	17876.534		
19	17873.284	39	17874.913	59	17876.612		
20	17873.340	40	17874.970	60	17876.655		

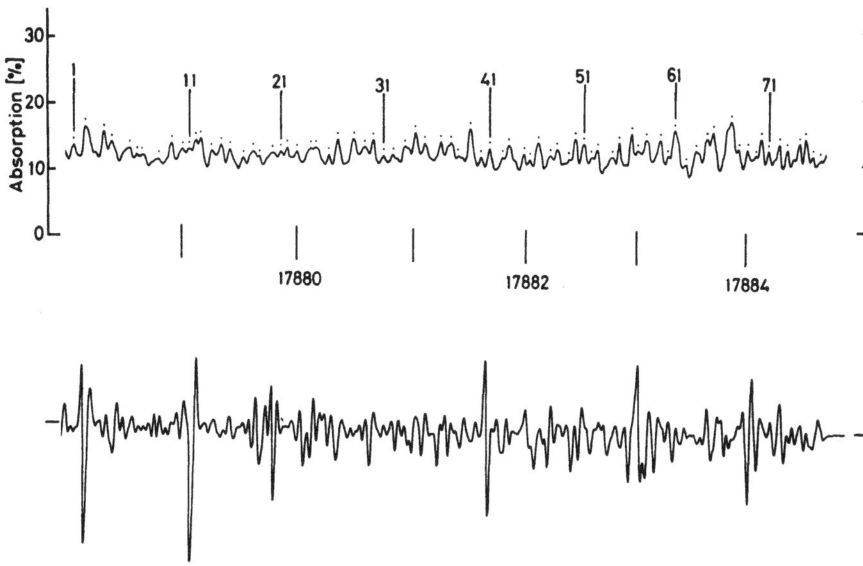

1	17878.097	21	17879.853	41	17881.677	61	17883.347
2	17878.199	22	17879.909	42	17881.791	62	17883.443
3	17878.343	23	17879.998	43	17881.847	63	17883.544
4	17878.405	24	17880.115	44	17881.978	64	17883.642
5	17878.555	25	17880.164	45	17882.035	65	17883.701
6	17878.613	26	17880.270	46	17882.108	66	17883.864
7	17878.668	27	17880.349	47	17882.225	67	17883.930
8	17878.805	28	17880.477	48	17882.272	68	17884.015
9	17878.913	29	17880.577	49	17882.382	69	17884.087
10	17879.005	30	17880.644	50	17882.434	70	17884.132
11	17879.063	31	17880.737	51	17882.507	71	17884.207
12	17879.114	32	17880.814	52	17882.582	72	17884.307
13	17879.157	33	17880.920	53	17882.632	73	17884.386
14	17879.250	34	17881.013	54	17882.773	74	17884.495
15	17879.328	35	17881.098	55	17882.831	75	17884.558
16	17879.404	36	17881.242	56	17882.949	76	17884.624
17	17879.528	37	17881.334	57	17883.003	77	17884.686
18	17879.619	38	17881.391	58	17883.084		
19	17879.664	39	17881.503	59	17883.207		
20	17879.760	40	17881.600	60	17883.280		

References

Chapter 1

1.1 D.K. Hsu, D.L. Monts, R.N. Zare: *Spectral Atlas of Nitrogen Dioxide 5530 to 6480* Å (Academic, New York 1978)
1.2 R.E. Smalley, L. Wharton, D.H. Levy: J. Chem. Phys. **63**, 4977 (1975)
1.3 T.D. Wilkerson, B. Ercoli, F.S. Tomkins: University of Maryland Technical Note BN-784(1974)
1.4 L.S. Rothman, A. Goldman, J.R. Gillis, R.H. Tipping, L.R. Brown, J.S. Margolis, A.G. Maki, L.D.G. Young: Appl. Opt. **20**, 1323 (1981)
1.5 L.S. Rothman, R.R. Gamache, A. Barbe, A. Goldman, J.R. Gillis, L.R. Brown, R.A. Toth, J.-M. Flaud, C. Camy-Peyret: Appl. Opt. **22**, 2247 (1983)

Chapter 2

2.1 D.K. Hsu, D.L. Monts, R.N. Zare: *Spectral Atlas of Nitrogen Dioxide 5530 to 6480* Å (Academic, New York 1978)
2.2 K. Uehara, K. Shimoda: Jpn. J. Appl. Phys. **16**, 633 (1977)
2.3 K. Uehara, H. Sasada, T. Kasuya: To be published
2.4 G.H. Williams, J.L. Hobart, T.F. Johnston, Jr.: "A 10-THz Scan Range Dye Laser", in *Laser Spectroscopy VI*, ed. by H.P. Weber, W. Lüthy, Springer Ser. Opt. Sci., Vol. 40 (Springer, Berlin, Heidelberg 1983) p. 422
2.5 T.A. Littlefield, D.T. Turnbull: Proc. Roy. Soc. (London) **A218**, 577 (1953)
2.6 A.D. Buckingham, D.A. Ramsey: J. Chem. Phys. **42**, 3721 (1965)
2.7 K. Uehara: Opt. Lett. **6**, 191 (1981)
2.8 K. Uehara: "Stark Modulation Spectroscopy in the Visible Region", in *Laser Spectroscopy V*, ed. by A.R.W. McKellar, T. Oka, B.P. Stoicheff, Springer Ser. Opt. Sci., Vol.30 (Springer, Berlin, Heidelberg 1981) p. 89
2.9 R.S. Mulliken: Can. J. Chem. **36**, 10 (1958)
2.10 J.B. Coon, F.A. Cesani, F.P. Huberman: J. Chem. Phys. **52**, 1647 (1970)
2.11 R. Solarz, D.H. Levy, K. Abe, R.F. Curl: J. Chem. Phys. **60**, 1158 (1974)
2.12 J. Hinze, R. Solarz, D.H. Levy: Chem. Phys. Lett. **25**, 284 (1974)

2.13 M. Krauss, R.J. Celotta, S.R. Mielczarek, C.E. Kuyatt: Chem. Phys. Lett. **27**, 285 (1974)
2.14 C.E. Jackels, E.R. Davidson: J. Chem. Phys. **64**, 2908 (1976)
2.15 C.G. Stevens, R.N. Zare: J. Mol. Spectrosc. **56**, 167 (1975)
2.16 T. Tanaka, R.W. Field, D.O. Harris: J. Mol. Spectrosc. **56**, 188 (1975)
2.17 J.C.D. Brand, W.H. Chan, K.A. Davis, J.L. Hardwick: J. Mol. Spectrosc. **54**, 74 (1975)
2.18 R.E. Smalley, L. Wharton, D.H. Levy: J. Chem. Phys. **63**, 4977 (1975)
2.19 K. Uehara: To be published
2.20 F. Bylicki, H.G. Weber: Chem. Phys. **70**, 299 (1982)

Chapter 3

3.1 W. Demtröder: *Laser Spectroscopy*, Springer Ser. Chem. Phys., Vol. 5 (Springer, Berlin, Heidelberg 1982)
3.2 H.P. Weber, W. Lüthy (eds.): *Laser Spectroscopy VI*, Proc. 6th Int. Conf., Interlaken, Switzerland, June 27-July 1, 1983; Springer Ser. Opt. Sci., Vol. 40 (Springer, Berlin, Heidelberg 1983)
3.3 A.R.W. McKellar, T. Oka, B.P. Stoicheff (eds.): *Laser Spectroscopy V*, Proc. 5th Int. Conf., Jasper Park Lodge, Alberta, Canada, June 29-July 3, 1981; Springer Ser. Opt. Sci., Vol. 30 (Springer, Berlin, Heidelberg 1981)
3.4 K. Sakurai, H.P. Broida: J. Chem. Phys. **50**, 2404 (1969)
3.5 K. Abe: J. Mol. Spectrosc. **48**, 395 (1973)
3.6 T. Tanaka, K. Abe, R.F. Curl, Jr.: J. Mol. Spectrosc. **49**, 310 (1974)
3.7 K. Abe, F. Myers, T.K. McCubbin, Jr., S.R. Polo: J. Mol. Spectrosc. **50**, 413 (1974)
3.8 H.D. Bist, J.C.D. Brand: J. Mol. Spectrosc. **62**, 60 (1976)
3.9 J.L. Hardwick: J. Mol. Spectrosc. **66**, 248 (1977)
3.10 R. Schmiedl, I.R. Bonilla, F. Paech, W. Demtröder: J. Mol. Spectrosc. **68**, 236 (1977)
3.11 J.C. Brand, J.L. Hardwick, R.J. Pirkle, C.J. Seliskar: Can. J. Phys. **51**, 2184 (1973)
3.12 G.I. Senum, S.E. Schwartz: Chem. Phys. Lett. **32**, 569 (1975)
3.13 C.G. Stevens, M.W. Swagel, R. Wallace, R.N. Zare: Chem. Phys. Lett. **18**, 465 (1973)
3.14 T. Tanaka, R.W. Field, D.O. Harris: J. Mol. Spectrosc. **56**, 188 (1975)
3.15 C.G. Stevens, R.N. Zare: J. Mol. Spectrosc. **56**, 167 (1975)
3.16 Y. Haas, P.L. Houston, J.H. Clark, C.B. Moore, H. Rosen, P. Robrish: J. Chem. Phys. **63**, 4195 (1975)
3.17 T.C. Lee, E.R. Peck: J. Mol. Spectrosc. **65**, 249 (1977)
3.18 J.C.D. Brand, K.J. Cross, A.R. Hoy: Can. J. Phys. **57**, 428 (1979)
3.19 D.L. Monts, B. Soep, R.N. Zare: J. Mol. Spectrosc. **77**, 402 (1979)
3.20 J.C.D. Brand, P.H. Chiu, A.R. Hoy: Can. J. Phys. **57**, 828 (1979)

3.21 H.D. Bist, J.C.D. Brand, A.R. Hoy: Can. J. Phys. **55**, 1453 (1977)
3.22 C. di Lauro: J. Mol. Spectrosc. **51**, 356 (1974)
3.23 J.C.D. Brand, P.H. Chiu: J. Mol. Spectrosc. **75**, 1 (1979)
3.24 M.J. Armstrong, J.C.D. Brand, C. di Lauro: Can. J. Phys. **59**, 559 (1981)
3.25 I. Raitt, A.M. Griffiths, P.A. Freedman: Chem. Phys. Lett. **77**, 433 (1981)
3.26 D. Neuberger, A.B.F. Duncan: J. Chem. Phys. **22**, 1963 (1954)
3.27 P.B. Sackett, J.T. Yardley: Chem. Phys. Lett. **9**, 612 (1971)
3.28 R. Solarz, D.H. Levy: J. Chem. Phys. **60**, 842 (1974)
3.29 F. Paech, R. Schmiedl, W. Demtröder: J. Chem. Phys. **63**, 4369 (1975)
3.30 V.M. Donnelly, F. Kaufman: J. Chem. Phys. **69**, 1456 (1978)
3.31 C.H. Chen, S.D. Kramer, D.W. Clark, M.G. Payne: Chem. Phys. Lett. **65**, 419 (1979)
3.32 T. Imasaka, T. Ogawa, N. Ishibashi: Chem. Lett. 405 (1977)
3.33 V.M. Donnelly, D.G. Keil, F. Kaufman: J. Chem. Phys. **71**, 659 (1979)
3.34 D.G. Keil, V.M. Donnelly, F. Kaufman: J. Chem. Phys. **73**, 1514 (1980)
3.35 S. Takezawa, N. Sugimoto, N. Takeuchi: Jpn. J. Appl. Phys. **21**, 809 (1982)
3.36 N. Sugimoto, S. Takezawa, N. Takeuchi: Jpn. J. Appl. Phys. **21**, 1536 (1982)
3.37 H.-J. Foth, H.J. Vedder, W. Demtröder: J. Mol. Spectrosc. **88**, 109 (1981)
3.38 W. Demtröder, F. Paech, R. Schmiedl: Chem. Phys. Lett. **26**, 381 (1974)
3.39 R.E. Smalley, B.L. Ramakrishna, D.H. Levy, L. Wharton: J. Chem. Phys. **61**, 4363 (1974)
3.40 R.E. Smalley, L. Wharton, D.H. Levy: J. Chem. Phys. **63**, 4977 (1975)
3.41 R.E. Smalley, L. Wharton, D.H. Levy: Acc. Chem. Res. **10**, 139 (1977)
3.42 D.K. Hsu, D.L. Monts, R.N. Zare: *Spectral Atlas of Nitrogen Dioxide 5530 to 6480 Å* (Academic, New York 1978)
3.43 H.J. Vedder, M. Schwarz, H.-J. Foth, W. Demtröder: J. Mol. Spectrosc. **97**, 92 (1983)
3.44 H.-J. Foth, H.J. Vedder: J. Mol. Spectrosc. **102**, 148 (1983)
3.45 T. Imasaka, T. Ogawa, N. Ishibashi: J. Chem. Phys. **70**, 881 (1979)
3.46 S. Takezawa, N. Sugimoto, N. Takeuchi: Chem. Phys. Lett. **97**, 77 (1983)
3.47 N. Sugimoto, N. Takeuchi, S. Takezawa: Chem. Phys. Lett. **99**, 475 (1983)
3.48 N. Sugimoto, S. Takezawa, N. Takeuchi: J. Mol. Spectrosc. **102**, 372 (1983)
3.49 C.H. Chen, D.W. Clark, M.G. Payne, S.D. Kramer: Opt. Commun. **32**, 391 (1980)

3.50 A.E. Douglas, K.-P. Huber: Can. J. Phys. **43**, 74 (1965)
3.51 M.J. Marsden, G.R. Bird: J. Chem. Phys. **59**, 2766 (1973)
3.52 G.R. Bird, M.J. Marsden: J. Mol. Spectrosc. **50**, 403 (1974)
3.53 H.D. Bist, J.C.D. Brand, R. Vasudev: J. Mol. Spectrosc. **66**, 399 (1977)
3.54 D.E. Tevault, L. Andrew: Spectrochim. Acta **30A**, 969 (1974)
3.55 F. Bolduan, H.J. Jodl: J. Mol. Spectrosc. **91**, 404 (1982)
3.56 H.J. Jodl, W.B. Holzapfel: Chem. Phys. Lett. **55**, 259 (1978)
3.57 D.M. Guthals, K.P. Gross, J.W. Nibler: J. Chem. Phys. **70**, 2393 (1979)
3.58 M.E. McIlwain, J.C. Hindman: J. Chem. Phys. **73**, 68 (1980)
3.59 K. Shimoda: "Double-Resonance Spectroscopy of Molecules by Means of Lasers", in *Laser Spectroscopy of Atoms and Molecules*, ed. by H. Walther, Topics Appl. Phys., Vol. 2 (Springer, Berlin, Heidelberg 1976) Chap. 3, pp. 197-252
3.60 T. Tanaka, A.D. English, R.W. Field, D.A. Jennings, D.O. Harris: J. Chem. Phys. **59**, 5217 (1973)
3.61 R. Solarz, D.H. Levy: J. Chem. Phys. **58**, 4026 (1973)
3.62 D. Solarz, D.H. Levy, K. Abe, R.F. Curl: J. Chem. Phys. **60**, 1158 (1974)
3.63 T. Tanaka, R.W. Field, D.O. Harris: J. Chem. Phys. **61**, 3401 (1974)
3.64 T. Tanaka: J. Mol. Spectrosc. **59**, 413 (1976)
3.65 R. Solarz, D.H. Levy: J. Mol. Spectrosc. **73**, 374 (1978)
3.66 I.P. Herman, A. Javan, R.W. Field: J. Chem. Phys. **68**, 2398 (1978)
3.67 B. Khoobehi, J.A. Roberts: J. Mol. Spectrosc. **97**, 227 (1983)
3.68 M. Allegrini, A.R.W. McKellar, P. Pinson, J.M. Brown: J. Chem. Phys. **73**, 6086 (1980)
3.69 J.M. Brown, T.C. Steimle, M.E. Coles, R.F. Curl, Jr.: J. Chem. Phys. **74**, 3668 (1981)
3.70 H.G. Weber, P.J. Brucat, W. Demtröder, R.N. Zare: J. Mol. Spectrosc. **75**, 58 (1979)
3.71 H.G. Weber, P.J. Brucat, R.N. Zare: Chem. Phys. Lett. **63**, 217 (1979)
3.72 F. Bylicki, H.G. Weber: Phys. Lett. **82A**, 456 (1981)
3.73 F. Bylicki, H.G. Weber: Chem. Phys. Lett. **79**, 355 (1981)
3.74 F. Bylicki, H.G. Weber: Chem. Phys. Lett. **79**, 517 (1981)
3.75 F. Bylicki, F. König, H.G. Weber: Chem. Phys. Lett. **88**, 142 (1982)
3.76 H.G. Weber, F. Bylicki, F. König, H. Burkhardt, J. Wieland: Z. Phys. A **309**, 363 (1983)
3.77 F. Bylicki, H.G. Weber: J. Chem. Phys. **78**, 2899 (1983)
3.78 H. Figger, D.L. Monts, R.N. Zare: J. Mol. Spectrosc. **68**, 388 (1977)
3.79 I.R. Bonilla, W. Demtröder: Chem. Phys. Lett. **53**, 223 (1978)
3.80 H.G. Weber, P.J. Brucat, R.N. Zare: Chem. Phys. Lett. **60**, 179 (1979)

3.81 H.G. Weber: Z. Phys. A **293**, 203 (1979)
3.82 M. Krauss, R.J. Celotta, S.R. Mielczarek, C.E. Kuyatt: Chem. Phys. Lett. **27**, 285 (1974)
3.83 C.F. Jackels, E.R. Davidson: J. Chem. Phys. **64**, 2908 (1976)
3.84 P.J. Brucat, R.N. Zare: J. Chem. Phys. **78**, 100 (1983)
3.85 J.-M. Flaud, C. Camy-Peyret, V. Malathy Devi, Palash P. Das, K. Narahari Rao: J. Mol. Spectrosc. **84**, 234 (1980)
3.86 S.P. Reddy, W. Ivancic, V. Malathy Devi, A. Baldacci, K. Narahari Rao, A.W. Mantz, R.S. Eng: Appl. Opt. **18**, 1350 (1979)
3.87 V. Malathy Devi, Palash P. Das, A. Bano, K. Narahari Rao, J.-M. Flaud, C. Camy-Peyret, J.P. Chevillard: J. Mol. Spectrosc. **88**, 251 (1981)
3.88 V. Malathy Devi, B. Fridovich, G.D. Jones, D.G.S. Snyder, Palash P. Das, J.-M. Flaud, C. Camy-Peyret, K. Narahari Rao: J. Mol. Spectrosc. **93**, 179 (1982)
3.89 V. Malathy Devi, B. Fridovich, G.D. Jones, D.G.S. Snyder, A. Neuendorffer: Appl. Opt. **21**, 1537 (1982)
3.90 J. Reid, M. El-Sherbiny, B.K. Garside, E.A. Ballik: Appl. Opt. **19**, 3349 (1980)
3.91 W. Dillenschneider, R.F. Curl, Jr.: J. Mol. Spectrosc. **99**, 87 (1983)
3.92 K.M. Evenson, H.P. Broida, J.S. Wells, R.J. Mahler, M. Mizushima: Phys. Rev. Lett. **21**, 1038 (1968)
3.93 R.F. Curl, Jr., K.M. Evenson, J.S. Wells: J. Chem. Phys. **56**, 5143 (1972)
3.94 S.M. Freund, J.T. Hougen, W.J. Lafferty: Can. J. Phys. **53**, 1929 (1975)
3.95 K. Hakuta, H. Uehara: J. Mol. Spectrosc. **94**, 126 (1982)
3.96 V.M. Baev, T.P. Belikova, M.B. Ippolitov, E.A. Sviridenkov, A.F. Suchkov: Opt. Spectrosk. **45**, 58 (1978) [English transl.: Opt. Spectrosc. **45**, 31 (1978)]
3.97 A.M. Angus, E.E. Marinero, M.J. Colles: Opt. Commun. **14**, 223 (1975)
3.98 N. Ioli, P. Violino, M. Meucci: J. Phys. E **12**, 168 (1979)
3.99 R.E. Teets, R. Feinberg, T.W. Hänsch, A.L. Schawlow: Phys. Rev. Lett. **37**, 683 (1976)
3.100 R.E. Teets, N.W. Carlson, A.L. Schawlow: J. Mol. Spectrosc. **78**, 415 (1979)
3.101 J.C.D. Brand, K.J. Cross, N.P. Ernsting: Chem. Phys. **59**, 405 (1981)
3.102 J.C.D. Brand, K.J. Cross, N.P. Ernsting, A.B. Yamashita: Opt. Commun. **37**, 178 (1981)
3.103 J.C.D. Brand, N.P. Ernsting: J. Mol. Spectrosc. **91**, 389 (1982)
3.104 D. Feldman: Opt. Commun. **29**, 67 (1979)
3.105 T. Suzuki: Opt. Commun. **38**, 364 (1981)
3.106 C.R. Webster, R.T. Menzies: J. Chem. Phys. **78**, 2121 (1983)

3.107 I. Burak, J. Tsao, Y. Prior, E. Yablonovitch: Chem. Phys. Lett. **68**, 31 (1979)
3.108 D. Feldmann, H. Zacharias, K.H. Welge: Chem. Phys. Lett. **69**, 466 (1980)
3.109 R.J.S. Morrison, B.H. Rockney, E.R. Grant: J. Chem. Phys. **75**, 2643 (1981)
3.110 R.V. Hodges, L.C. Lee, J.T. Moseley: Int. J. Mass Spectrom. Ion Phys. **39**, 133 (1981)
3.111 G. Radhakrishnan, D. Ng, R.C. Estler: Chem. Phys. Lett. **84**, 260 (1981)
3.112 I. Tanaka: J. Photochem. **17**, 487 (1981)
3.113 V.S. Antonov: Opt. Spectrosk. **52**, 10 (1982) [English transl.: Opt. Spectrosc. **52**, 5 (1982)]
3.114 R.J.S. Morrison, E.R. Grant: J. Chem. Phys. **77**, 5994 (1982)
3.115 C.B. McKendrick, C. Fotakis, R.J. Donovan: J. Photochem. **20**, 175 (1982)
3.116 Y. Matsumi, Y. Murasawa, K. Obi, I. Tanaka: Laser Chem. **1**, 113 (1983)

Chapter 4

4.1 D.K. Hsu, D.L. Monts, R.N. Zare: *Spectral Atlas of Nitrogen Dioxide 5530 to 6480 Å* (Academic, New York 1978)

Subject Index

FSC
www.fsc.org

MIX
Papier aus verantwortungsvollen Quellen
Paper from responsible sources
FSC® C105338

If you have any concerns about our products,
you can contact us on
ProductSafety@springernature.com

In case Publisher is established outside the EU,
the EU authorized representative is:
Springer Nature Customer Service Center GmbH
Europaplatz 3, 69115 Heidelberg, Germany

Printed by Libri Plureos GmbH
in Hamburg, Germany